ROMANTIC NEOCLASSICISM

浪漫新古典 VI

欧朋文化 策划
黄滢 马勇 主编

華中科技大學出版社
http://www.hustp.com
中国 · 武汉

创意新古典
CREATIVE NEOCLASSICISM

传统新古典
TRADITIONAL NEOCLASSICISM

创意新古典

雨后塞纳河

Seine After Rain

- 项目名称：南通中海 · 上林院北区 275 别墅样板房
- 项目位置：江苏南通
- 设计公司：上海桂睿诗建筑设计有限公司
- 设计师：桂峥嵘、张晓薇、冯磊、张艳玲、李成祥
- 摄影师：钱达
- 用材：雕刻白大理石、碳化木地板、进口壁纸
- 面积：522 m^2

- Project Name: No. 275 Show Villa, CSCL (Nantong)
- Location: Nantong, Jiangsu
- Design Company: G&K International Design Institution
- Designer: Grace Kwai, Zhang Xiaowei, Feng Lei, Zhang Yanling, Li Chengxiang
- Photographer: Qian Da
- Material: White Marble, Carbonized Wood Flooring, Wallpaper
- Area: 522 m^2

如果你是一个浪漫梦幻的人，那么，你一定会喜欢本案的设计。当你走进它，你会不自觉地想起雨后的清晨，漫步在塞纳河畔，远眺巴黎圣母院桥，古老的河畔带来无限的震撼与遐想。是的，这就是设计师想告诉我们的故事，他们以塞纳河与雨后的天青色为引子，将塞纳河的妩媚迷人和天青色的典雅带入这套样板房作品中，结合浪漫的法式装饰风格，试图营造一种神秘优雅的美感。

本案定位在较为女性化的如同雨后的天青色以及梦幻般的空间感受，在简洁清爽的灰蓝调子基础上，辅以闪亮时尚的钻石切割元素，具有钻石切割造型的镜面，搭配具有丰富遮光感的材料，营造出如同珠宝般令人目眩神迷的视觉感受，烘托家的绚丽。

If you are a romantic person, you are supposed to like the interior design of this project. Once within, you unconsciously think of a morning after rain, when you walk along Seine, overlooking Notre Dame de Paris in the distance and wandering your mind here and there. That is what's this project is aimed for. A cyan Seine now serves as the introduction to take in the charm and the cyan grace of the river into to make an exploration into an elegant mystery in combining romantic French style.

Positioned more morbidezza, like the cyan after rain to allow for dreamy feelings, this project is decorated with cut-diamond elements on the basis of gray-blue tone. The cut-diamond mirror not only enriches light-shadow materials, but also exerts a jewelry-dazzling vision to highlight a gorgeous home.

铜雀春深，岁月流金

As Spring Comes Deep into Copper Sparrow Platform, Time Becomes Good

- 项目名称：世茂集团苏州铜雀台样板间项目
- 项目位置：江苏苏州
- 设计公司：上海李孙建筑设计咨询有限公司
- 设计总监：邢政
- 设计师：范帅锋
- 软装设计：陈小玲
- 摄影师：陈盛
- 用材：米黄色石材、高光漆、橘色硬包、手绘壁纸、玫瑰金钛金条
- 面积：510 m^2

- Project Name: Copper Sparrow Show Villa, Shimao Group (Suzhou)
- Location: Suzhou, Jiangsu
- Design Company: JSP Architects Ltd.
- Director Designer: Xing Zheng
- Designer: Fan Shuaifeng
- Upholstering Designer: Chen Xiaoling
- Photographer: Chen Sheng
- Materials: Marble, Glossy Paint, Hard Bumps, Hand-Painted Wallpaper, Rose Gold Tantalized Bar
- Area: 510 m^2

建筑的真正价值，在于居住于其间的人，在于光阴和故事。因此，在本案苏州铜雀台别墅样板间的内部，设计打破了中西风格的界限，以人的生活为本营造空间。从“一个人”、“一群人”、“一家人”这三种主要的家庭情感活动需求出发，划分私人层、家庭层、私享层三种类别空间，以人为出发点，实现家族的冷暖交汇。

恢宏的别墅样板房空间，在软装设计方面则表现为法式清新风格。设计师在传统的法式上做了简化，保留传统元素，在材质及表现手法上融入现代的处理，搭配爱马仕(Hermès)的经典元素，使空间显得更为灵动。雕花的样式、精致的线条、精雕细琢的家具、精美的布艺、随处可见的花艺以及其他的浪漫元素，比比皆是，营造出业主对待生活悠然自得的美好画面，任岁月流金，仍美如童话。

设计师苛刻甄选顶级的天然石材，更恪守雕刻装饰传统，力求使空间于方寸之间皆近乎完美。细部的钛金条手法，在经典中彰显精致，恰如其分地表达了主人对高品质生活的热爱。而绚丽、明亮且层次丰富的色彩，又使本案给人留下了深刻的印象。经典大气的爱马仕橘色，时尚不减的黑白色、华贵辉煌的金色、纤巧梦幻的水蓝色、清新自然的绿色、美丽静谧的湖蓝色……丰富的颜色组合，使空间俨如一幅色彩斑斓的风情画卷，在设计师的精心编织与描绘下，引人入胜。加上家具的陈设与精致饰品的布置，如结合爱马仕经典元素图案打造的墙面、造型简洁而独特的精致沙发、晶莹剔透的水晶吊灯、法式优雅而富有情调的地毯等，营造出豪华高贵的空间氛围。

A building in its real sense focuses on the occupants, time and stories. This makes a show flat for villas which has broken away the boundary eastern and western, human-oriented to create differentiated floors for individuals, a group and a whole family, of private, family and enjoyment to achieve a harmony under the same roof.

Inside the magnificent interior is a French style, fresh but simplified to have kept the traditional elements in fusing materials and techniques with modern treatment and classical Hermes elements to etherealize the space. Besides carving, lines, furnishings and fabrics, everywhere are romantic items and floriculture to reveal a casual attitude toward life and that as times goes good, life is still as good as that in fairy tale.

Natural stones involved are meticulously selected, cut or carved scrupulously by tradition, so any inch in the space almost remains in an impeccable state. Titanium bar highlights the delicacy out of the classical, duly and properly expressing personal love and passion for life. Gorgeous and bright, colors are quite impression. Hermes orange, black-white, gold, water blue, green, and acid blue altogether make a multi-colored picture scroll. So attractive and enchanting it is to make a luxurious and noble ambience with furniture and accessories, like Hermes-patterned wall, concise but unique sofa, chandelier and graceful, appealing carpet.

JAYA
CONTEMPORARY DESIGN
WITH A PEDIGREE

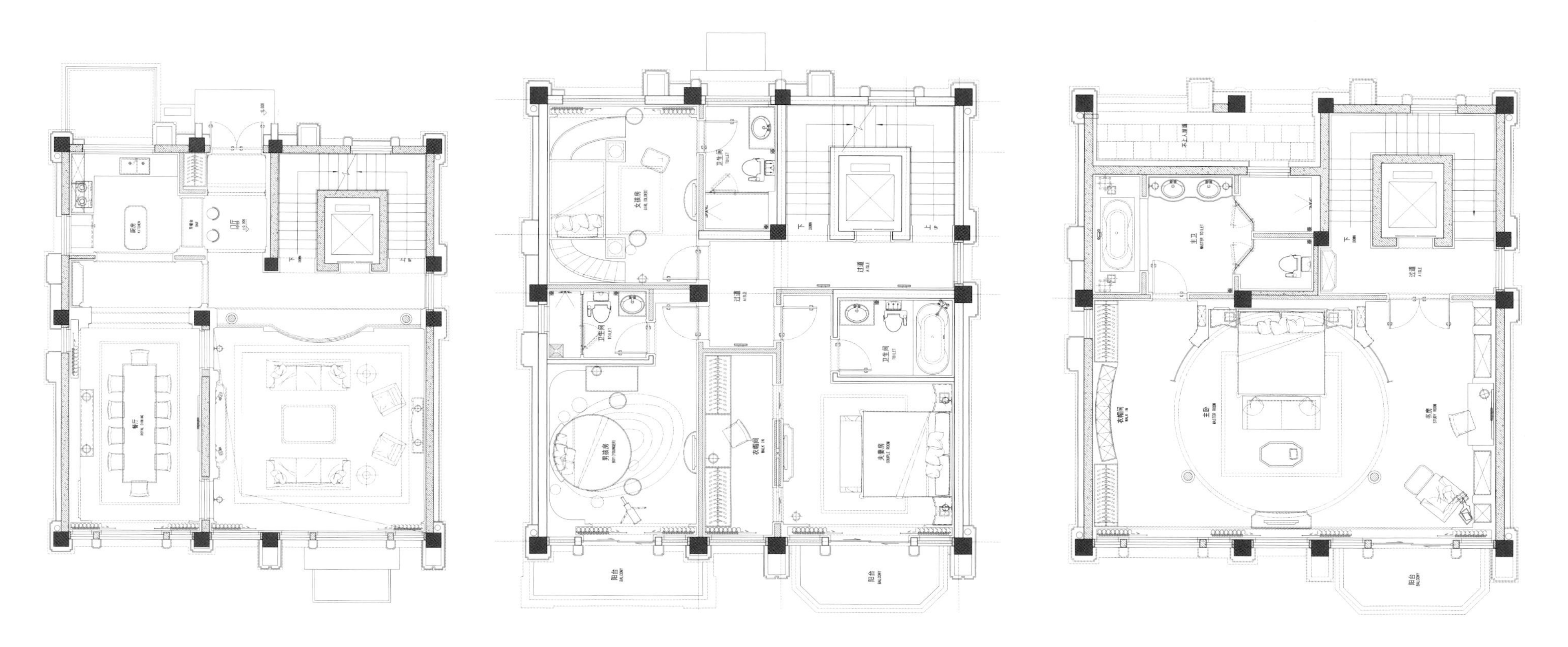

纤美如诗，一曲意大利清新小调

An Italian Minor: As Fine as a Poem

■ 项目名称：意大利卡普里蒂贝宫酒店　　■ Project Name: Capri Tiberio Palace

卡普里蒂贝宫酒店（ Capri Tiberio Palace ）坐落于 Faraglion 岛上一栋始建于 19 世纪的建筑内，距离卡普里岛（ Capri ）的中央广场（ Piazzetta) 步行仅 2 分钟。卡普里蒂贝宫酒店拥有让人叹为观止的美景。酒店的室内设计让人过目不忘：当然室内设计师 Giampiero Panepinto 也功不可没，其设计灵感来源于 20 世纪 50—70 年代间流行的美学元素。该酒店共有客房 60 间，客房内的陈设，包括色彩、壁纸、织物等均由 Limonta 与 Dadar 品牌为该酒店定制。

流连于叶子花、龙舌兰等众多花卉掩映下的 Taschen 书店内，绝对是阅读爱好者的人生一大乐事，艺术、建筑及电影方面的书籍这里应有尽有。酒店还拥有 1 个全景屋顶露台。大部分的客房设有 1 个阳台或露台，部分客房享有海景。所有客房都配有空调、迷你吧和免费无线网络连接。餐厅供应当地的地中海美食，客人可以在海景露台上用餐，而犹太菜肴在另一个区域供应。现代风格的 Jacky Bar 拥有别致的装饰。卡普里蒂贝宫酒店设有带土耳其浴室、桑拿浴室和热水浴缸的健身中心，并提供按摩服务和美容理疗服务。

Located in a building of the 1800s on the island of Faraglion, Capri Tiberio Palace is 2 minutes' walk to the central plaza of Piazzetta. A project it is that has a landscape that's acclaimed as the peak of perfections and an interior design by Giampiero Panepinto that's bound to be unforgettable. All inspiration takes roots in the aesthetics popular between the 50s and 70s. All 60 guest rooms are fixed with furnishings and accessories of Limonta and Dada brand are customized, including hues, wallpapers and fabrics.

Hidden in vegetables, flowers and maguey is the bookshop of Taschen, absolutely a life fortune for those bookworms because a wide range of books on art, construction and films are available. The terrace on the roof is panoramic. When most of the rooms take direct access to one balcony or terrace, and some enjoys a view of sea scape, all are embellished with air conditioning, mini bar and Wi-Fi.

The restaurant has Mediterranean cuisine. Terraces of sea scape make good places for dinning. As for Jew dishes, they are served in special area. Jacky Bar is of modern style and embellished with unique decoration. Additionally, there are bathroom of Turkey and sauna, fitting center with hot-water bathtub, and service of massage and physiotherapy.

PAN AM
TRAVELOGUES

PAN AM

ACQUA PANNA

L 英式贵族生活礼遇

ife of British Noble

- 项目名称：上海安缦西郊英式样板房
- 设计公司：上海无相室内设计工程有限公司
- 设计师：王兵、王建
- 软装设计：李欣
- 摄影：三像摄 / 张静
- 用材：樱桃木、钛金板、皮革软包、石材、玻璃马赛克、真丝壁布
- 面积：400 m^2

- Project Name: Aman British Show Flat, West Suburb, Shanghai
- Design Company: Shanghai Wuxiang Interior Design Engineering Co., Ltd.
- Designer: Wang Bing, Wang Jian
- Upholstering Design: Li Xin
- Photography: Threeimage / Zhang Jing
- Materials: Cherry, Titanium Board, Leather, Marble, Glass Mosaic, Silk Wallpaper
- Area: 400 m^2

本案以昔日的英国风格类比成功人士在事业上的巅峰，以英国历史为记叙主线，通过室内的空间细节及丰富的陈设来精心营造一种典雅庄重，兼具绅士风度的住宅空间。客厅空间以米色为主，金属与温润雅致的石材整合起来，与咖啡色系的花纹布艺沙发、古典丝织地毯，华丽的波西米亚水晶灯具形成强烈的碰撞，将欧洲社会的整体美学精神呈现出来。转入家庭室，一股浓浓的南亚帕米尔高原及印度联邦气息迎面而来，仿佛置身于一场宴会之中。透过早餐厅的门又被中国传统特色的手绘真丝墙纸的精致所吸引，真是应了那句“吃在中国”的俗语。不经意间已经来到书房，空间陈设上的狩猎文化，再一次将思想带到非洲原始森林，仿佛一时间猎枪声四起，鸟惊兽鸣。最后北美风格的客房风格给这段穿越历史时光的旅程画上了句号！空间整体风格强调英式贵族气质的融合，彰显出传统英式风格中独有的低调与高雅，让传统素养、严谨的做派在此套住宅中得到了完美的诠释。

It is a project where the style of British makes an analogy of those distinguished people who have reached their career peak. It is a space where achievements British ever made throughout the world now serve as the narrating principal. And it is a space where details and a wealth of furnishings and accessories altogether make a solemn elegance with gentle demeanor. In a setting of beige, the living room blends the use of metal and marble, where the coffee floral fabric sofa, the classical silk carpet, and the Bohemia chandelier make a sharp but intended collision to present a whole aesthetics exclusive to Europe. The family room features a strong air that comes across Pamirs and Indian Union, where you feel nothing but in a feast. Servants from foreign countries are busy around. Your eyesight pierces through the door of the breakfast room, and then is fixed onto the hand-made silk wallpaper. That's just like an old say goes, that China makes a good place for you to enjoy cuisine. In the study, furnishings and accessories are sealed with venery culture, leading your thought into African primitive forest, in which birds are scared into sky and animals are frightened to run around. The guest room is of North American style, where your journey through time tunnel comes to an end.

The combination of British noble into the whole highlights its unique characteristic low-key but elegant, so traditional cultivation and rigid approaches are endowed with a perfect interpretation.

一场袂影飘香的生活派对

Life Party with Sleeve and Shadow Becoming Fragrant

- 项目名称：绿地赵巷海珀风华别墅
- 设计公司：集艾室内设计（上海）有限公司
- 设计师：黄全
- 参与设计：方建、赖智
- 摄影：三像摄 / 张静
- 面积：350 m²

- Project Name: The Show Villa of Sea Ample, Greenland (Shanghai)
- Design Company: G&A Interior Design (Shanghai) Co., Ltd.
- Designer: Huang Quan
- Participant: Fang Jian, Lai Zhi
- Photography: Threeimages / Zhang Jing
- Area: 350 m²

室内设计引入“跨界”时尚设计理念，将奢侈品牌融入室内，彰显独具品位的客户群，国际化的品质感，继承与改良以及不断创新。线条元素的提取及细部金色勾勒、高雅灰色基调的运用等赋予了室内高贵、优雅、时尚的品质感。

“挑空”已经成了别墅、豪宅的代名词，它足以表达室内的气派、奢华、高贵。“开放式西厨设计的引入，一场西方生活方式的演绎”不仅是西方装饰手法的应用，是西方时尚前卫生活方式的体验，也是与客户所向往的西方生活品位的融合，更是跨界理念的升华。

A project this space makes where to take in concepts of fashion for implanting within luxury brands, a field-cross approach with which to stand for a consumer group with special taste, a global quality, and heritage, improvement and innovation. With line elements, detailed gold sketch, and gray hue, the interior is endowed with sense of nobility, grace and fashion.

"Hollowed-out" has now been the synonymous of villa and mansion, for its connotation is enough to bring the temperament, luxury and nobility out of the space, where the western kitchen is readily available as a token not only to interpret a western lifestyle to offer a leading life experience, but combine a fusion of life taste of the West in sublimating the field-cross concept.

DAGGER
OF THE
MIND

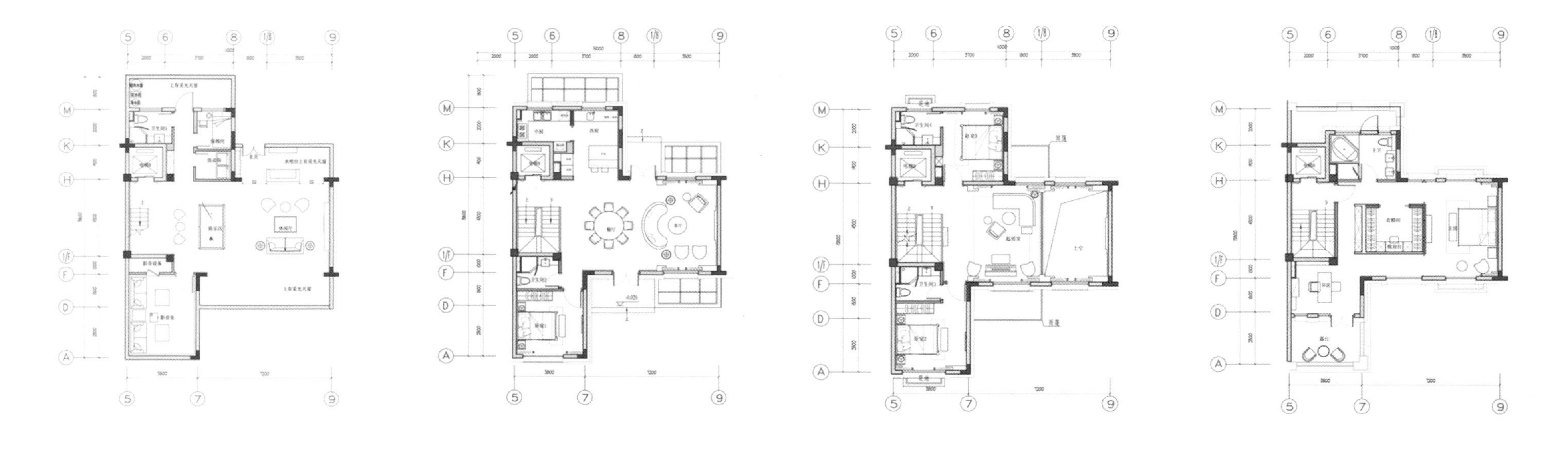

RIGHT ROYAL

B低调高雅，英伦贵族

British Aristocracy: Reserved but Elegant

- 项目名称：沈北新湖仙林金谷英式样板房
- 设计公司：上海无相室内设计工程有限公司
- 设计师：王兵、王建
- 摄影：三像摄 / 张静
- 用材：樱桃木、钛金板、皮革软包、石材、玻璃马赛克、真丝壁布
- 面积：400 m²

- Project Name: British Show Flat
- Design Company: Shanghai Wuxiang Interior Design Engineering Co., Ltd.
- Designer: Wang Bing, Wang Jian
- Photography: Threeimages / Zhang Jing
- Materials: Cherry, Titanium Board, Leather, Marble, Glass Mosaic, Silk Wallpaper
- Area: 400 m²

室内设计以英伦风格打造出的是一个典雅庄重，兼具绅士风度的住宅空间。传统素养、严谨的风格在此套住宅中得到了完美的诠释。为配合定位，在整体格调上以极尽奢华的姿态，迎合客户品位。理想的空间尺度，华丽的灯光布置，典雅雍容的气质是打造此尊贵项目的先决条件。

步入室内，空间颜色以米色为主，一条主线贯穿玄关、厨房、餐厅、客厅。深色木墙与米黄色皮革软包相互搭配，局部墙面采用灰水鸭蓝和橄榄绿色的搭配。整体色调以典雅的灰色调为主，局部点缀高贵的暗红、矿紫、亮金和水鸭蓝色。富贵典雅，强调英式贵族气质的融入，对于造型设计及材质配搭，设计师选用低调奢华的手法将高反射度的玻璃、金属与温润雅致的石材整合起来，并与咖啡色系的花纹布艺、沙发、地毯形成碰撞，丰富了室内的生活气息，沉稳而不古板，在灯光配合下，彰显出传统英式风格中独有的低调与高雅。

British style create an elegance and gentlmanly space. Traditional and rigid approaches and aesthetics are interpreted in an impeccable way. The holistic luxury abuts with the position while gorgeous light fixture, grace and magnificence accomplishes prerequisite in order to meet customer's taste in a space with ideal dimension.

In a setting dominated with cream-colored hue, a traffic line goes throughout the vestibule, the kitchen, the dining room, and the living room. Dark wooden walls coordinate with beige leather upholstering, while parts of some walls are coated in gray-duck-blue and olive green. Gray in the whole, patches are adorned with dark red, mine-purple, bright gold and gray-duck-blue. The rich, and the elegant are stressed with the air of British nobility. With low-key but luxurious approaches, highly-reflective glass and metal are blended with warm marble in colliding with the coffee patterned fabric, sofa, and carpet. The life atmosphere is thus enriched, staid and equally lovely to bring out the unique refined constraint with light cast on.

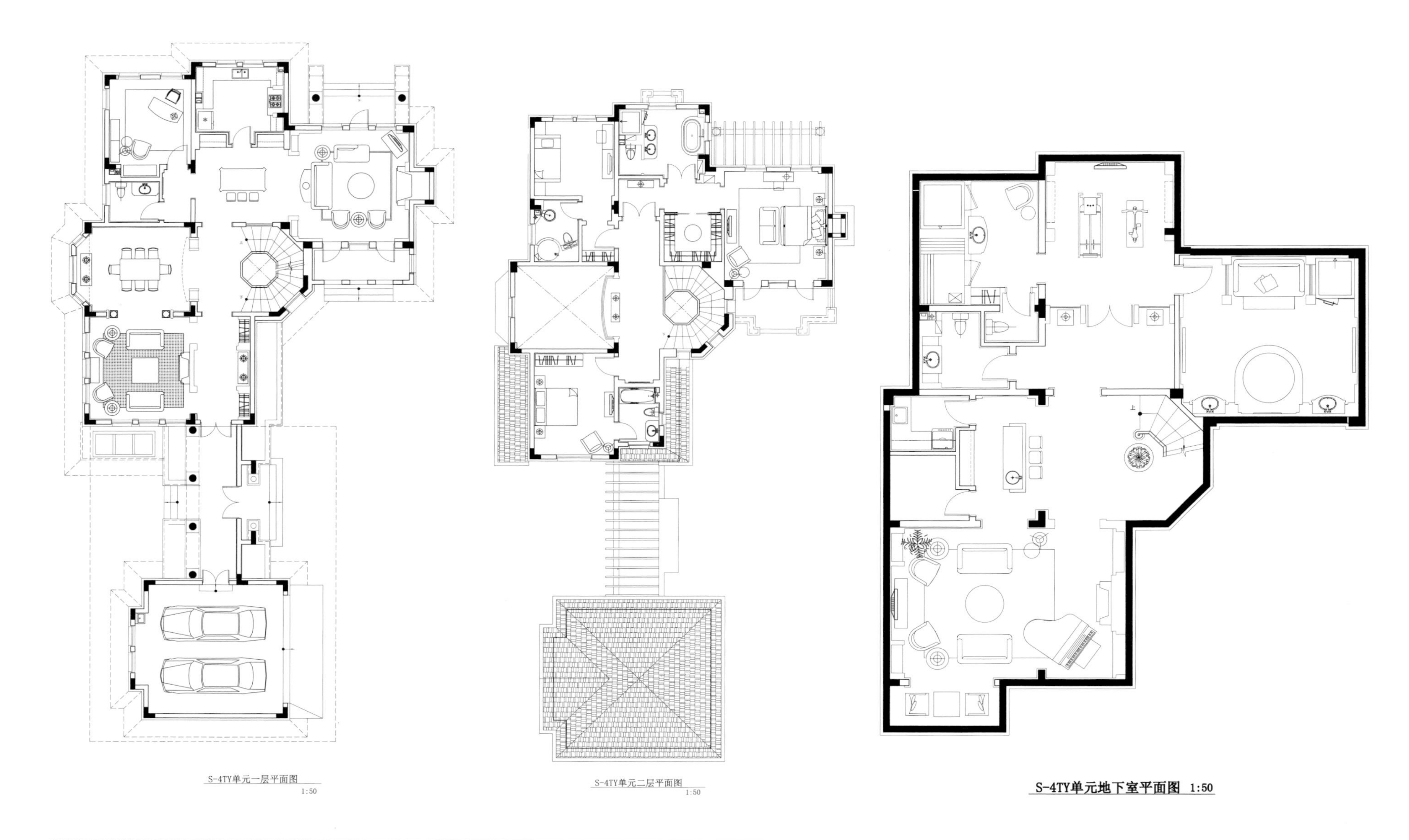
S-4TY单元一层平面图
1:50
S-4TY单元二层平面图
1:50
S-4TY单元地下室平面图 1:50

为了爱马仕的不羁和自由

For Hermes's Unruly Freedom

- 项目名称：爱马仕
- 项目位置：重庆永川
- 设计公司：品辰设计
- 设计师：夏婷婷
- 参与设计：余丽华
- 用材：皮革、金属、石材
- 面积：242 m²

- Project Name: Hermes
- Location: Yongchuan, Chongqing
- Design Company: Pinchen Design
- Designer: Xia Tingting
- Participant: Yu Lihua
- Materials: Leather, Metal, Marble
- Area: 242 m²

人生的真谛在于不断攀登一个又一个高峰，
达成目标，然后马不停蹄地奔向下一个目标，
不求俯瞰天下，悦心即可。
这是一个时尚而浓烈的时代，
橙色和金色的搭配华贵而不浮夸，
各种皮革与野兽元素的采用凸显了一种野性的美感。
旅行的自由和马背上的不羁，
看似复古却演绎着奢华的贵族生活。
世界在变，初心不变，
狂野不在表面，
而在永不停步的内心，
强大在绅士的外表和深厚的底蕴，
唯有低调沉稳的绅士底蕴让高雅空间延续。

ART

THE GLENLIVET
15

The true meaning of life lies when one peak has been overcome, another one is right in front waiting for you, and when one aim has been accomplished, you are eager to start the next one. Be realistic not to stand ahead of the world. A contented heart is what's desired. This is an age fashionable and strong, when the match of orange and gold is not showy, for leather of all kinds and animal elements can highlight a wild aesthetics. Like freedom in journey or that on horseback, the retro is virtually makes a vivid interpretation of life exclusive to nobility. Though a world it is that has been changing, the real intention remains as it is without any alternation. The wild in its real sense cannot be made on surface, but into heart or behind a genteel appearance and a profound background. So only with a low-profile, staid and gentle interior, can the elegance and grace in space be continued.

W梦幻空间旋舞曲

Whirling Dance of Illustrative Dream

- 项目名称：兰尼勒公馆
- 设计公司：金斯顿拉弗蒂设计
- 设计师：拉弗蒂、金斯顿
- 摄影师：墨菲

- Project Name: Ranelagh Residence
- Design Company: Kingston Lafferty Design
- Designer: Roisin Lafferty, Susanna Kingston
- Photographer: Donal Murphy

“兰尼勒公馆”原本的结构并不完整，出入受限。18 世纪的建筑静静地立于都柏林的一隅，除了地下室，两层的空间几近处于坍塌的境地。如此先天的条件限制对于本案设计而言可谓是难度倍增。

数年无人问津，对于建筑结构的完全保持是毁灭性的打击。修复时，本着从建筑本身及其周围环境的保护出发，设计新添了众多结构性的部件。旧时的建筑面积虽然开阔，但平面上及后花园的使用上却是极其有限。对建筑面积充分地使用，尽可能地引入自然光线是本案设计的主要目标。因此，整个地下室空间予以降低，提高了上层空间的“地空”距离。

室内空间及后花园采取“深挖” 的手法，原有的建筑因此得到了最大化的保护。同时，花园分层设计的采用，有利于更多的自然光线进入地下室。修复后的空间，彰显现代。但线性的造型、原生的用材，升华着建筑固有的特点，映衬着空间传统的风格。

Ranelagh Residence was structurally unsound, conservation protected with highly restricted access, the development of this mid terrace, Georgian Town-house remains our most challenging project to date. Located on a quiet square in one of Dublin's most sought after residential postcodes, this 18th century, 2 storeys over basement property was in a dilapidated condition.

VAN GOGH

The property had been neglected for a number of years, which subsequently proved detrimental to the properties structural integrity. Extensive structural works were required to support both the existing and proposed structures and to retain the neighboring properties. Although the property has a generous floor area, the existing layout and rear garden were negatively impacting the usability of the space. The main objective of the design was to maximize the space available and direct as much natural light as possible into the existing property. The entire basement floor level was reduced to improve floor to ceiling. Height and the rear garden was extensively excavated to allow for the construction of a large conservatory extension. A tiered garden was designed to increase the quality of natural light entering the basement level. The linear forms and raw materials create a contemporary space which juxtaposes beautifully within the period features of the original property and the traditional style conservatory.

FROM BRIEF TO BUILD
TURNER

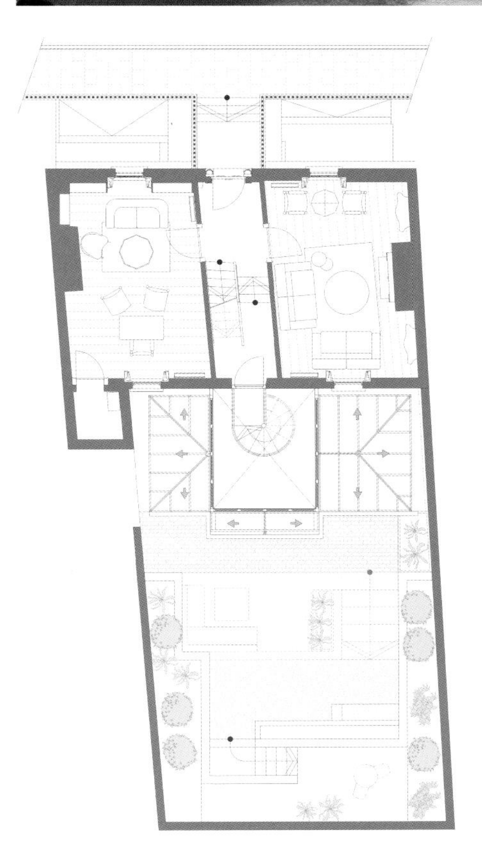

W 无限灵感，艺术如花般绽放

With Inspiration, Art in Blossom like Flowers

■ 项目名称：迈阿密黑蝴蝶酒店 ■ Project Name: Viceroy Miami

迈阿密“黑蝴蝶”酒店其实是个奢华的度假村。东方风韵的室内设计，国际化标准的服务，独特的氛围成就了本案无与伦比的地位。

148 间客房、套房尽情满足了极具甄别眼光的国际游客。一居室、两居室的套房要么配有全套的厨具，要么是全设施的厨房。纵情于各处空间，酒店提供的是名副其实的度假体验。

15 楼，泳池旁的餐厅除了鸡尾酒，还有精致盘碟佳肴、大厨亲手制作的酒店招牌菜。24 小时营业的客房餐饮，满足了客人的不同需要。站在阳台，凭栏四望，风景分外妖娆。

该精品酒店位于布里克尔大街北边的门户，周围遍布别致小区、文化机构、体育场地及旅游设施，对于各位商界精英来讲，真是极为便利。“超级 50 酒廊”高居酒店 50 楼，可以说是迈阿密最与众不同的酒廊。带有酒吧的私人泳池，其风景在迈阿密难觅第二。

公园之旁，海湾之滨的位置设有舞台，便于开展商务会议、新闻发布会、摄影展、促销、产品发布等活动。多功能空间，如传统的会议室、招牌酒店、天台酒廊、Spa 等让本案成了一个宜集会、宜聚会的良好场所。

多种服务可满足如婚宴、成人礼等各种活动。富有创意的概念美食、事务策划等满足了客人对装饰、设计、视听、多媒体等方面的需要。

Viceroy Miami blends timeless East Asian accents, international service standards and Miami's own rich ambience to create an unparalleled resort destination in Brickell, Miami's Wall Street of the Americas.

Created for discerning globetrotters, this luxury resort in Miami offers 148 residentially appointed guestrooms, suites and one- and two-bedroom residences (which include fully equipped kitchenettes or full kitchens), and more non-negotiable indulgences for a truly exceptional Miami resort experience.

From cocktails and small plates served poolside on the 15th floor, to the chef's tasting menu at the signature restaurant and terrace overlooking the pool deck, to round-the-clock guestroom dining, Viceroy Miami satisfies every culinary craving.

Located at the northern gateway to Brickell Avenue, Viceroy Miami is ideally situated to cater to the business elite. This Miami boutique hotel is also centrally located near the city's many distinctive neighborhoods, cultural institutions, sporting venues and area attractions. Perched atop the 50th floor of Viceroy Miami, FIFTY Ultra Lounge offers the city's

most exclusive lounge, while the private pool deck with bar offers dramatic skyline views unequalled in Miami.

At Viceroy Miami, set the stage bayside and park-side for a business meeting, press event, photo shoot, sales presentation, conference or product launch. Versatile function spaces – ranging from traditional meeting rooms to the signature restaurant, rooftop lounge, and The Spa at Viceroy Miami – make Viceroy Miami an ideal location for any gathering or meeting.

The hotel also offers comprehensive catering services for every event from weddings to bar mitzvahs - including innovative culinary concepts and event planners that can assist you with all your event needs from decor, design, audio/visual or media.

收藏生活艺术
Collection of Life Art

- 项目名称：乡村别墅
- 设计公司：基里尔室内设计装潢
- 设计师：基里尔

- Project Name: Country House
- Design Company: Kirill Istomin Interior Design & Decoration
- Designer: Kirill Istomin

年轻夫妻的传统家居以象牙白、淡褐色的调色板作为背景。大大的客厅内，丝绸质感的帷幔搭配极富喜感的斑马条纹。女主人的房间内，装饰风格的墙纸给人一种旧时好莱坞的感觉。全白的厨房妆点着几何瓷砖，极具张力，时髦而又实用。

几乎所有房间均可直通阳台，乡村生活的重要元素尽在其中，室内外空间也因此融为一体。除了卧室，其他空间配备着单色、无图案的亮光窗帘，升华着空间乡村风格的质感。

各功能空间全部位于一楼。玄关旁边有一个书房与冬日花园相连。这里是主人最喜爱的地方。卡迪拉克车载座椅应主人喜好扶手椅的形式出现在本案空间。只不过椅子上新蒙上了深色的皮革，并搭有面料的边穗，别致的搭配削弱了扶手椅的沉重质感。同样厚重质感的窗户因为竹质百叶、经典窗帘等的使用也变得轻盈透亮。

最富有魅力的空间当是与餐厅毗邻的客厅。长长的厅式空间开有一排法式大窗。曾经的泳池如今已然不在。餐厅、客厅之间的界限也已刻意模糊。紧靠着客厅、餐厅的是厨房，主人对烹饪的喜爱尽在其中。沙发后的镜面墙反射着透过窗户的光，放大着此处的空间视觉感。包括主人房、儿童房在内的隐私区域全部位于二楼，均配有浴室、化妆室。三楼还另外设有书房、客厅。花卉墙纸让客房如厅轩一般。客厅的枕头、窗帘都绘有同样的图案。同时此处也是本案唯一一个色调轻盈明亮的空间。

“乡村别墅”是为了日常生活起居使用，而不是为了周末度假。着色安静、轻松，轻装定制。家具配饰新旧杂陈。铁艺桌更是美国传奇名流布鲁克·阿斯特曾经使用过的物品。

An ivory and ecru palette provides a fresh backdrop for a young family's traditional furnishings. Istomin carefully curates all patterns: in the large living-dining room, the silk curtains are edged with a jazzy zebra while in the master boudoir, a Deco-style wallpaper channels old Hollywood. A dramatic geometric tile in the all-white kitchen is both stylish and practical.

Almost every room in this house has access to the terrace. It is one of the most important elements of life in the countryside. It helps to feel connection between the interiors and the nature. Istomin tried to keep this feeling by creating monochrome light curtains without any prints in every room except bedrooms.

Common areas are situated on the first floor. There's a study with connected winter garden next to the entrance. It is client's favorite place in the house. Once he mentioned that wants armchairs as in Cadillac – and Kirill made them in deep Cogniac shaded leather with fabric fringe on the periphery. This unusual combination allowed armchairs look less heavy. Bamboo blinds, included in the composition together with classic curtains and lambrequins, also make the windows look less heavy.

The most glamorous space in the house is the living room adjacent to the dining room. It's a long hall with a row of French windows. It used to be a swimming pool but clients decided to rebuild it. The border between the living and the dining rooms is almost non-existent. Next to them is the kitchen – it is clear from the first look that inhabitants of this house love to cook. Mirror wall behind the sofa reflects the light from the windows and makes this space look even bigger. Private rooms – master bedroom and kid's room, each with its own bathroom and dressing room, is situated on the second floor. On the third floor there's another study and guest bedroom. Walls are covered with flower printed wallpapers that makes the guest bedroom look like a pavilion. The same print is on the pillows and curtains. It is the only room with bright decor in the house.

This house is for everyday living, not for weekends. That's why it was so important to keep the colors calm and relaxed. All the upholstery is tailor-made. As usual, Istomin uses antique furniture and mixes it with new pieces. The most interesting object here is an iron table that belongs to legendary American socialite Brooke Astor.

W 以奢华艺术，奠定法式新古典的非凡荣耀

With Artistic Luxury to Make Glory of the French Neoclassicism

■ 项目名称：巴黎半岛酒店 ■ Project Name: The Peninsula, Paris

巴黎半岛酒店是半岛酒店集团进军欧洲市场的首个项目，奠定当代设计及豪华舒适住宿新标准。酒店大楼为超过百年的古典建筑，经过复修翻新，并装配最现代化设施，重现华丽气派，书写半岛传奇新篇章。

巴黎半岛酒店位于市中心地段克勒贝尔大街 19 号 (Avenue Kléber)，位置无比优越，与凯旋门等举世闻名的名胜古迹、博物馆及高级购物区仅举步之遥。

巴黎半岛酒店大楼是一座 19 世纪末法式古典风格的酒店建筑，原址是 1908 年开幕的当时巴黎最有代表性的大酒店之一。酒店大楼别具法式古典建筑特色，带有新古典美学神韵，为 19 世纪奥斯曼男爵 (Haussmann) 改造巴黎市容后，极为普遍的建筑风格。经大规模翻新复修的巴黎半岛酒店，配备 21 世纪现代化设施；设有 200 间客房，其中 34 间为套间，其豪华舒适与住宿享受皆属巴黎顶尖之列。

酒店翻新工程历时 6 年，云集法国最优秀的工匠，其大楼外墙、名贵大理石、华丽木壁板， 以及金箔等建筑及装潢细节均至臻美善，尽得法式精湛工艺精髓。

The Peninsula in Paris claims its existence of the brand hotel in European, accomplishing new standard of luxurious, cotemporary and comfortable accommodation. Though refurbished from a classical building that's over 100 years, it's modern-facilitated, presenting a new chapter for Peninsula.

Its location in Avenue Kléber, the primary place in Downtown is really incomparable, when adjacent to Arch of Triumph and steps' away from other world-famous historical sites, museums and high-end precincts.

Its construction, one of the most representative hotel then that opened in 1908, is a neoclassical French building at the end of 19th century. A project it is done when Haussmann popularized construction style in the improvement of city's appearance. After being renovated to a large extent, here has modern facilities with 200 guest rooms including 34 suites. It's ranked top in terms of whether extravagance or comfort.

The recondition process takes craftsmen across France 6 years. Construction and decoration involves facade, rare marble, gorgeous wood siding and gold foil. All are no doubt intended for an utmost beauty and endowed with the quintessence of French craft.

S浮光掠影，上海之梦

Skimming over the Surface, Dreaming in Old Shanghai

- 项目名称：上海华贸 · 东滩花园项目 H1 户型
- 项目位置：中国上海
- 设计公司：上海壹陈建筑装饰设计工程有限公司
- 设计师：陈峰
- 用材：橡木染色、香槟银箔漆、地毯、啡慕斯大理石、白玉兰大理石、艺术墙纸、银镜
- 面积：336 m^2

- Project Name: Unit H1, East Bund Garden
- Location: Shanghai, China
- Design Company: Shanghai Yichen Decoration Design Engineering Co., Ltd.
- Designer: Chen Feng
- Materials: Dyed Oak, Champagne Paint, Carpet, Marble, Wallpaper, Mirror
- Area: 336 m^2

每个人心中都有一个上海！海派风格对于每一个在上海生活的人而言，都有不同的理解。但有些方面是共通的，譬如说小资情调，譬如说浮华与暧昧，譬如说东西结合的包容性。对于设计师而言，海派风格是一种浪漫而戏剧化的气质。本案想要呈现的是一个老上海明星的家。因为只有这个主题，才跟海派风格才是匹配的。

所以，我们看到的是犹如电影画面一般的场景感。陈逸飞的海上系列，老唱机里放着夜上海的歌曲，地下室里陈列着各种各样的旧唱片和主人的收藏品，照片墙上主人年轻而充满光彩的脸庞。

这是一场关于浮光掠影的“上海之梦”。

In everyone's heart, there is a city of Shanghai. For anyone living in such a city, honored as Magic Capital, the style of Shanghai has numerous different perception. But something common is certain, that is petty bourgeoisie taste, showy ambiguity and tolerance. For designers, Shanghai style is more like a temperament romantic and dramatic. So a dwelling place for a star in old Shanghai is intended in this place. Only with such a theme, can the style of Shanghai be matched.

Scenes like those in movie comes into eyesight naturally. Works of the series of Shanghai are by Chen Yifei, songs on old Shanghai sound from old player, all kinds of old record and personal collection are displayed in the basement, and on the picture wall are the owner in his youth with vigor and shinning face.

That is all about the dream of Shanghai, skimming over the surface.

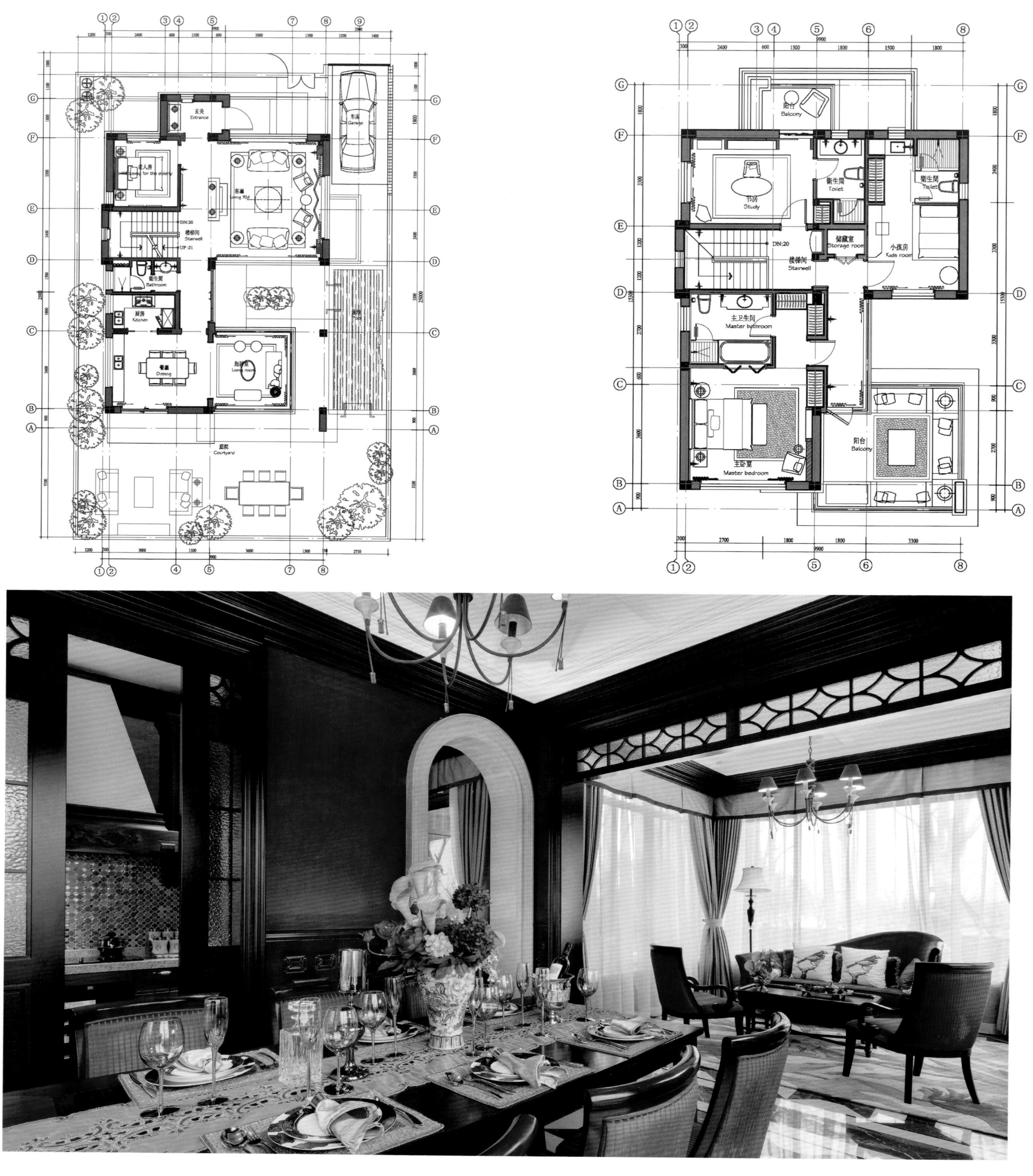

玄关
Entrance
车库
Garage
老人房
Room for the elderly
客厅
Living RM
DN:20
楼梯间
Stairwell
UP:21
卫生间
Bathroom
厨房
Kitchen
餐厅
Dining
Living room
庭院
Courtyard
阳台
Balcony
书房
Study
卫生间
Toilet
卫生间
Toilet
储藏室
Storage room
小孩房
Kids room
DN:20
楼梯间
Stairwell
主卫生间
Master bathroom
主卧室
Master bedroom
阳台
Balcony

卫生间
Bathroom
工作间
Workplace
吧台
Bar
酒窖
Living room
保姆间
Nanny between
楼梯间
Stairwell
品酒区
Tasting area
UP 20
收藏室
Favorite Room
视听室
Audio-visual room

R优雅国度，醉人香颂

Realm of Grace: Enchanting Fragrance

- 项目名称：常州新城帝景法式别墅
- 项目位置：江苏常州
- 设计公司：上海桂睿诗建筑设计咨询有限公司
- 设计师：桂峥嵘
- 摄影师：徐盛珉
- 面积：648 m²

- Project Name: French Villa, Imperial Landscape
- Location: Changzhou, Jiangsu
- Design Company: G&K International Design Institution
- Designer: Grace Kwai
- Photographer: Xu Shengmin
- Area: 648 m²

本案是一个三代之家的常住空间，设计师考虑到了男女主人和其父母与子女的性格、爱好，并将这些因素融入设计表达中，满足个人自自由以及生活需求。设计师以成熟的表现手法，将法式风格理念植入空间设计中，找寻那些可以穿越历史与时空的美感和情怀表达的细节，意欲彰显一种经典而优雅的奢华。

空间构造上，本案呈现出精雕细琢的美感，且极富视觉层次感，特别是在地下一层的红酒雪茄区“拱形”门廊和窗户及雕花的天花设计，营造出十足的浪漫与梦幻气息。色彩上，呈现出绚烂夺目的视觉效果，同时用色大胆，充满着跳跃与灵动感。优雅的蓝色调依然是设计师青睐和擅用的色彩，使整体空间极富现代、时尚气息。在极富层次感的硬装基调下，本案在软装饰上也下足了功夫，质地上乘且融合古典与时尚美感的家具配置，华贵的布艺、优雅的水晶吊灯、工艺品摆件、装饰画等，以史诗般的恢弘气度及艺术化的组合方式，将家装扮成一个优雅的国度，在这里好像能欣赏到塞纳河左岸流淌着的悦耳香颂。

The French-styled villa is for a family with three generations living under the same roof, which naturally take various personality and hobbies to meet individual life demands. A project this space where to highlight a classical but elegant luxury by implanting French style to seek aesthetics expression that can go through time tunnel and details to express feelings. All is as usual, done with mature approaches.

Rich in a strong sense of layers, the interior brings out beauty, accomplished with careful effort and meticulous work. The arched porch on the first basement is for bar of red wine and cigar. With the window and carved ceiling, the porch brings forward a very strong sense of romantics and dream. The hues when shaping a gorgeous vision, is bold, jumping and lithe. The blue the designer is expert at using is complimentary to the modern and fashion. The holstering is paid equal attention. Furniture classical and fashion, magnificent fabric, elegant chandelier, art pieces and painting, all are grouped with epic manner, altogether transferring the space into elegance and grace. Here you feel as if you were listening to a chanson on left of Seine.

PARIS

Fires in the Dark

美式优雅下潜藏的野性不羁

Wide Nature Behind American Grace

- 项目名称：山东东营万方小区 C 户型样板间
- 设计公司：上海壹陈建筑装饰设计工程有限公司
- 设计师：李小斌
- 软装设计：张妍
- 摄影：三像摄 / 张静
- 用材：帝王石大理石、摩洛哥金啡大理石、米白洞石大理石、硬包、墙纸、实木复合地板
- 面积：269 m^2

- Project Name: Unit C Show Flat of All Places, Dongying, Shandong
- Design Company: Shanghai Yichen Decoration Design Engineering Co., Ltd.
- Designer: Li Xiaobin
- Upholstering Designer: Zhang Yan
- Photography: Threeimages / Zhang Jing
- Materials: Marble, Hard Bumps, Wallpaper, Lamination Flooring
- Area: 269 m^2

美剧的流行也带来了美国文化的盛行，引导着当下一代人的饮食、穿衣风格、生活态度，美式的文化也在家居装饰业中蔓延开来。

本案设计师打造的美式家居风格既有文化气息、贵族气质，又不缺乏自在与情调，这些元素也正迎合了时下的文化资产者对生活方式的需求。

浅色调的墙体与黑、暗红、褐色及其他深色的软装饰品形成有力的视觉冲击，同时映衬出空间的包容性。沉稳粗犷的深色家具，强调厚重与实用性，彰显迷人细节的造型，纹理、雕饰的细腻高贵，散发着亘古而久远的芬芳。

儿童房搭配出一个航海的梦想，无论从蓝白红配色上，还是轮船造型的床，充满海洋风情如锚造型的饰品，都在诠释着勇敢的美国探险精神。

设计师用空间陈设的语言讲述了当今城市中坚分子自然纯真的生活追求，以及其骨子里透着的些许野性不羁。

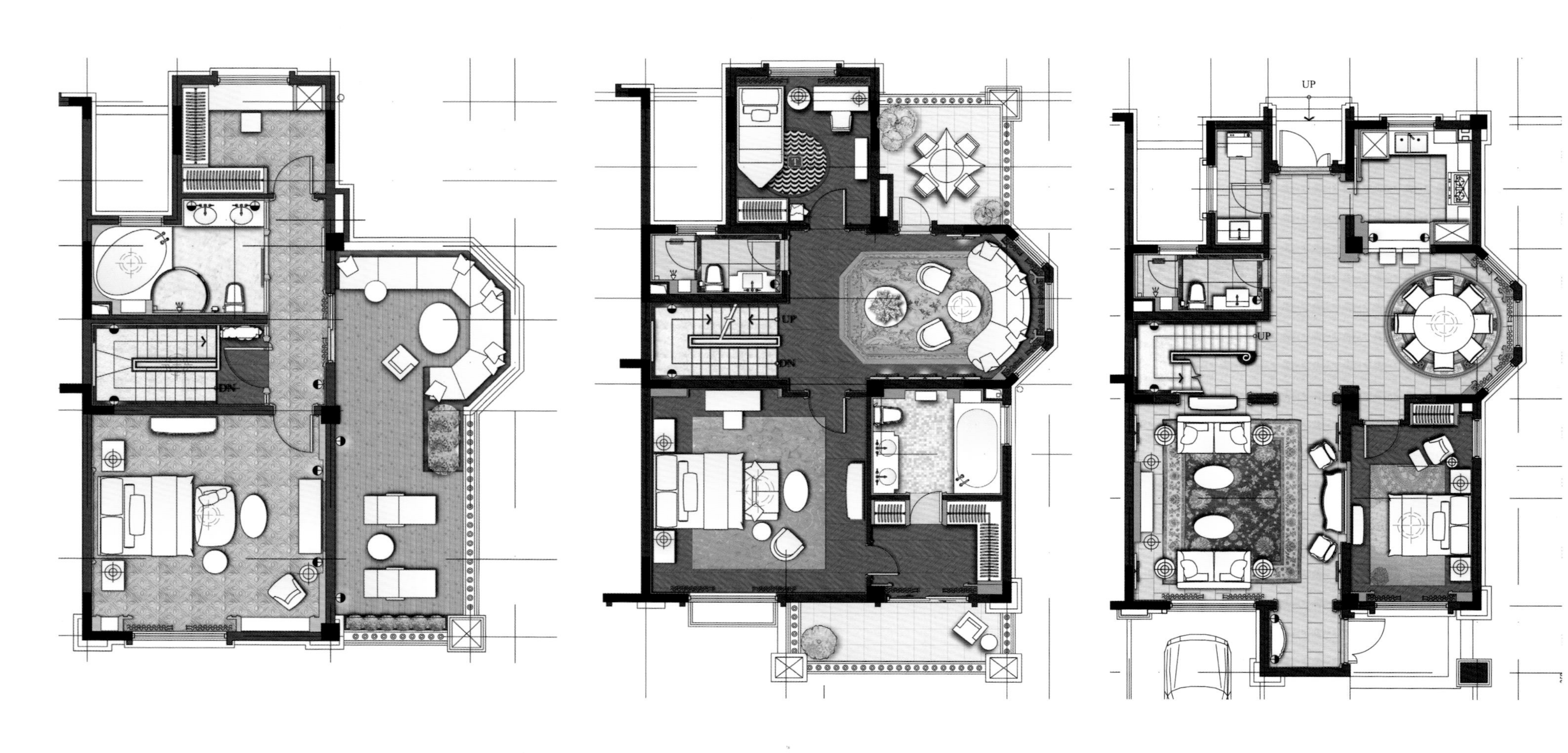
UP
UP
DN
UP
DN

With the popularity of American drama is the wide acceptance in the world of American culture. When leading the dinning, dressing and living of the young generation, American culture is overspread in interior decoration.

As for this project, its American style is not only cultural, noble but also free and sentimental. All are in good consistence with the life style preferred among those with money and education.

The light hues of the wall make a sharp and intended contrast with the black, the dark red, the brown and other dark colors of the upholstering while exerting strong visual impact and setting off the spatial tolerance. The rough but staid furniture stresses more on the historical sense and the practicability; the charming of pattern, space, sculpture is highlighted together with delicate and noble; and the fragrance is bound to last forever.

The room for the child brings out a navigation dream, where whether the match of blue, white and red, or the steamer-like bed, or the accessory of ocean style, interprets well the spirit to explore of America.

With spatial language, the pursuit in life of the urban hard core is depicted natural and innocent, yet equally wild and untamed inborn in the bone.

CAVS
23

高贵与优雅的碰撞：当空间设计遇上贵族蓝

Collision of Noble and Grace: When Space Design Meets the Blue of Nobility

- 项目名称：贵阳伯爵钻石广场样板房
- 设计公司：大阅艺术设计机构
- 设计师：吴喜丰

- Project Name: The Show Flat of Count Diamond Square, Guiyang
- Design Company: Dayue Art Design Agency
- Designer: Wu Xifeng

本次样板房设计由大阅艺术设计机构与贵州伯爵集团及资深室内设计师吴喜丰先生联手打造。设计尝试在装修与装置、加法与减法上进行取舍，在装修上崇尚减法，在装置上却推崇加法设计，一切从简，没有过多地使用风格符号。设计师致力于空间规划，使区域功能更具合理性与科学性。

推门而进，即被眼前这般丰富的景象所震撼，惊喜轮番上演。石材塑造的宽阔空间里，直线与曲线相互穿插，交替变化、融合，是理性和感性的完美结合。造型优美的家具、精致的饰品、多样的壁纸、丰富的图案……这些造型、质感、色调各不相同的珍贵物品，琳琅满目，丰富多彩，在设计师的巧妙安排下，协调共存，营造出一种亦中亦西的独特韵味。

代表着美丽、冷静、礼制、安详与广阔的蓝色，被运用于本案设计中，可谓是一个大胆而明智的尝试。全世界范围内，对于蓝色的评价非常高。在英国，贵族血统被称为“蓝血”，皇室和王族女性所穿的深蓝色服装被称为“皇室蓝”。在基督教中，蓝色是圣母玛利亚的象征。在伊斯兰教中，蓝色是一种纯洁的颜色。蓝色还象征着年轻，在中国都有“青春”（青即指蓝色）一词，是希望之色。据说，孔雀蓝是除了金色之外最难驾驭的色彩之一。

一抹又一抹的贵族蓝、孔雀蓝，化为地毯上美丽的图案、青花瓷、生动的挂画、柔软的窗帘。也许它是一种最冷的色彩，但在这里，它确实让整个空间变得更加出众，给人留下深刻的印象。它与金色结合，看似不搭调，却能营造出独具特色的高贵感。金色为它借出温暖的火光，提升了空间的温度，带来热情的跳跃感。而白色以及木色系的加入，则又巧妙地化解了两者之间的强强碰撞。

The show flat makes a trial to apply "minus" in decoration and "add" in device. Without overmuch stylish symbols, more attention paid to space plan leads to a reasonable and scientific function.

The interior strikes an immediate imposing impression. In a setting shape with marble, curves and straight lines are interwoven and interspersed to tell of an impeccable integration reasonable and emotional. Furniture, accessories, wallpaper and patterns spread floods of beautiful things into eyes, coordinate and harmonious to build up a flavor Eastern and Western.

Applied throughout is a bold but wise trial, blue, beautiful, calm, ritual, peaceful and broad. As it has a universal praise, the noble in Britain is honored as blue blood, and the dark blue for imperial females is called "royal blue". In Christian, blue is the symbol of Virgin Mary. In Islam, blue is a hue of purity. In China, blue means youth, a word literally pronounced as Qingchun in Chinese language, and Qing refers to blue. And peacock blue is alleged as the one most difficult to handle besides golden.

And it's here the color of peacock blue has been transferred onto carpet, China blue-white porcelain, hanging picture and curtain. Maybe it can make a coldest hue, but it becomes outstanding, unique, and stylish to leave a good impression. With golden, it seems not so suitable, but presents a special and unique sense of nobility, for it with golden, raises the spatial temperature in bringing forward jumping sense. The collision between, however, is offset because of white and wood.

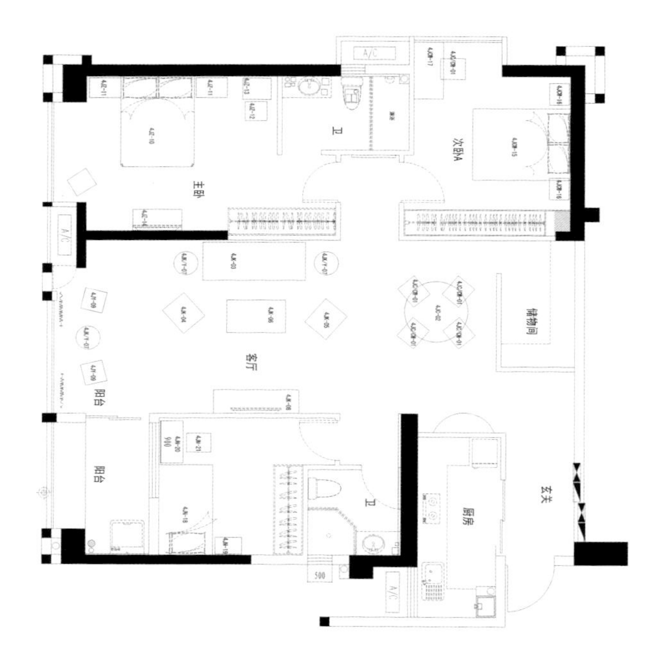
A/C
主卧
次卧A
客厅
储物间
阳台
厨房
玄关

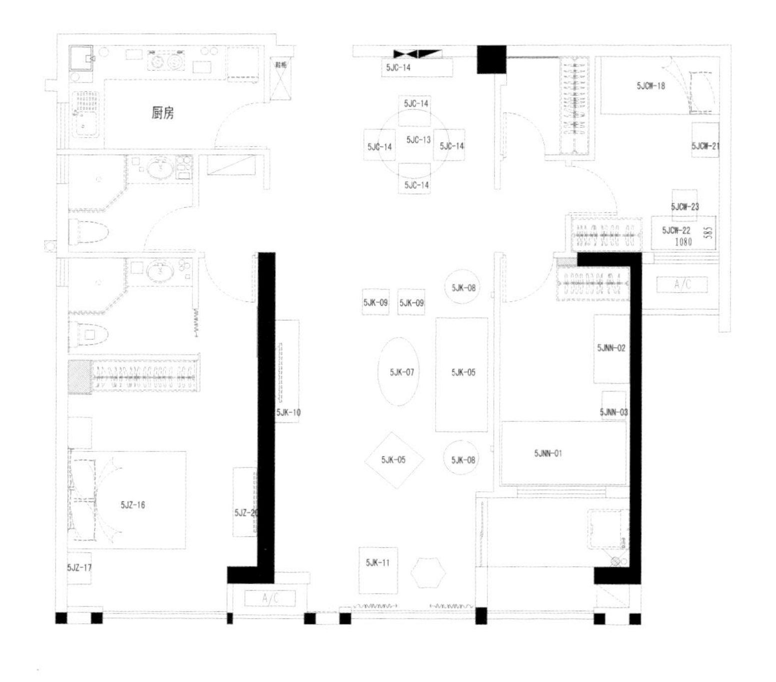
厨房
5JC-14
5JC-13
5JCM-18
5JCM-21
5JCM-23
5JCM-22
A/C
5JK-09
5JK-08
5JK-07
5JK-05
5JNN-02
5JNN-03
5JNN-01
5JK-10
5JK-05
5JZ-16
5JZ-20
5JZ-17
5JK-11
A/C

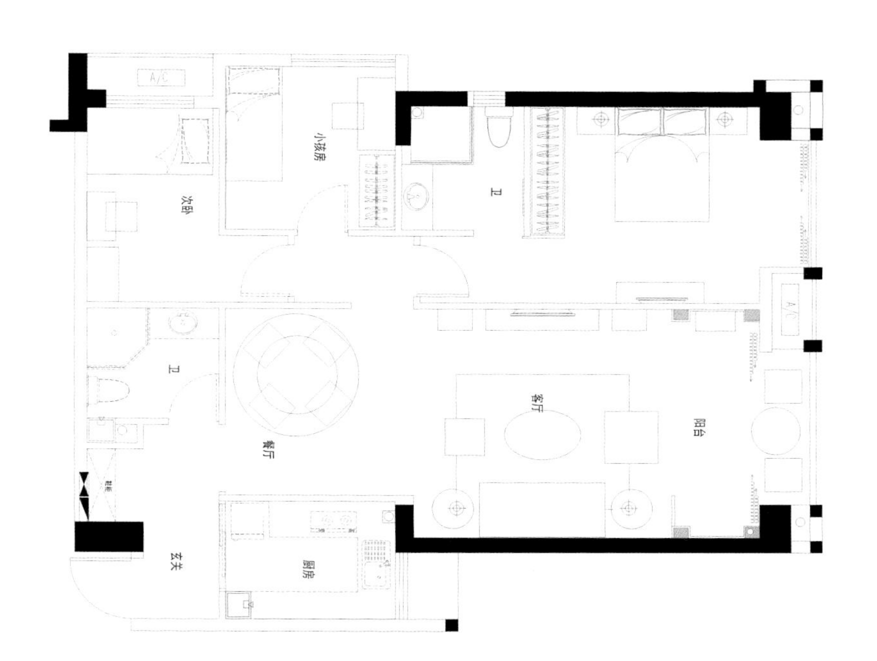

A/C
小孩房
次卧
卫
卫
A/C
客厅
阳台
餐厅
鞋柜
玄关
厨房

34

共生力量

Intergrowth

- 项目名称：永清生态庄园联排别墅
- 项目位置：河北廊坊
- 设计公司：RWD 设计师事务所
- 设计师：黄志达
- 参与设计：迈克 · 阿道夫
- 软装设计公司：Ambiance 软装设计团队
- 软装设计：Apple

- Project Name: Yongqing Ecological Townhouse
- Location: Langfang, Hebei
- Design Company: Ricky Wong Designer Ltd.
- Designer: Ricky Wong
- Participant: Mike Adolph
- Upholstering Design Company: Ambiance Design
- Upholstering Designer: Apple

在本案中，设计师关照居住者对自由与生活的追求，在设计与生活之间搭建一座桥梁，达到传统生活审美意境与现代生活方式的自由沟通联结，寻求西方形体与东方情韵的相互碰撞融合。设计着重尊贵有品位的生活追求，也蕴涵一种共生的生活态度。人与自然、设计与生活、整体与部分、东方与西方、传统与现代、新潮与复古、坚硬与柔软之间共生共存，演绎法式风格的浪漫与雅致。

生活追求的最好表现是室内设计，本案中设计师将米白色运用得淋漓尽致，简约干练，直抵人心，让人产生无限的遐想。流畅的几何线条演变为空间意识形态，勾勒出迷人的空间质感，让人随着或硬朗、或纤细的几何线条感知简约之美，顷刻间征服视线。设计师贯彻“Total Solution”全方位解决方案概念，从空间到陈设，整体服务专业化、精细化，考虑到了生活的全部及各种素材元素之间的共生，用品质建造惬意舒适的生态居所，正如其传达的理念：设计生活！

软装陈设的灵感源于生态城概念，软装设计师契合整体空间基调，兼容并蓄，配以精致的装饰和纹样，运用工笔画技艺及浮雕工艺等，将国风雅韵的元素以现代、时尚的西方形体呈现，多材质、多元素、多工艺巧妙并置，在空间发生奇妙的化学反应，塑造出东情西韵并融的特殊美感。一个以无色系为主的空间中采用点缀色的搭配可以造就一个清新脱俗的空间。本案中柠檬黄的抱枕与一些装饰罐形成一个围合线；孔雀绿的沙发、台灯、摆件又连成一条围合线，这个空间中绿色沙发与孔雀黄靠包成为两条色彩线的交汇点，是点缀色围合的交互中心。

餐厅空间里最引人注目的就是中间绿叶点缀的白黄色花艺，巧妙地将空间中黄、绿、白、青等主要颜色都集中在此。金色吊灯、绿色餐巾、金色餐巾扣与主花艺形成了主从关系，花艺成为无可替代的点睛主角。
空间中的色彩布局是有节奏的，尤其是色彩形状对节奏的影响更甚，异型色块构成动感的节奏，让人心情明快；规整色块构成慢节奏，令人心情平和。睡眠区慢节奏的床品将色彩引入画内。抽象的不规整芥末黄和孔雀绿色块点缀的地毯承载着整个卧室，让空间变得跳跃灵动，年轻人的青春、活力、奋斗与含蓄表达得淋漓尽致。

A project this space makes where to stress personal freedom quality and life pursuit, bridging design and life to communicate traditional aesthetics and modern life in accomplishing a collision between western form and structure and eastern charm. A life pursuit with taste implies a life attitude to realize intergrowth. Here tells of French romantics and elegance by combining people and nature, design and life, part and whole, east and west, modern and tradition, fashion and retro, and tough and soft.
The pursuit of life at its best is interior design. And throughout the space, the color of creamy white is used incisively and vividly, simple and concise to get a direct access into heart which on the contrary has abundant imagination. Flowing geometric lines have been refined and shifted into the spatial ideology, sketching a charming texture. Along lines soft or hard, delicate or geometric, vision are conquered all of a sudden. With the concept of "Total Solution" carried out, the intergrowth really brings forward ecological leisure and comfort, of space, furnishings and accessories, service, professionalization, refinement, life and materials. This is destined to be conveyed: to design life.
The inspiration of upholstering is rooted in eco-friendly city, by which to fit in with the holistic tone, incorporative and embellished with decoration and pattern, like technique of collaborate-style painting and process of relief to juxtapose Chinese tradition and modern western form and body with the use of materials, elements and process. Chemical changes take place in a same space to shape a special aesthetics that's of the East and the West.

A space free of heavy colors can make a fresh and refined project, once embellished with colors in proper amounts. Cushions of lemon yellow and decorative pots make an enclosure while peacock-green sofa, desk lamp and ornaments complete another. The green sofa and the peacock-yellow cushion are the conjunction of two color lines, or rather a mutual center of colors around.
The salient feature is the white-yellow flower with green leaves in its middle, as a finishing touch to be complimentary to the gold chandelier and napkin button, and green napkin.
Cadent is the color arrangement. The color shape exerts a great impact on spatial rhyme. Irregular color lump is dynamic to make mood higher, while the one regular is staid to calm heart. The bedding of slow rhyme introduces the hue within. Carpet of mustard and peacock green spreads over the space, which in turn becomes etherealized, by which the vigor and vitality of the youth is fully and completely presented with their diligence and connotation.

穿越时空的戏剧舞台

Dream Stage Through Time Tunnel

■ 项目名称：苏克酒店（法国巴黎）

■ Project Name: Hotel Maison Souquet

“苏克酒店”位于巴黎北区，繁华的蒙马特所在，典型的巴黎联排别墅风格。新颖的精品酒店彰显波希米亚的自由不羁。艺术家、贵族、上流人士、舞蹈家云集于此，蜂拥至剧场、酒吧、私人沙龙。20 世纪初，对于巴黎上流社会及美学大师而言，巴黎别墅如同第二个家。装饰奢华的“苏克酒店”，如今就为世人充当了这样一个地方。对蒙马特及蒙马特著名的歌舞厅“红磨坊”来讲，如同一块石子投入静静的水面，轻轻地泛着涟漪。

于巴黎的闹市抽身，“隐匿”于本案之中，享受着每一个房间的奢华。中央的楼梯、书房，集东方魅力与其他风情于一体。精致的护墙板，美国科尔瓦多的皮质家具呼应着折中风格的装饰物及原创的艺品。大厅里的鸡尾酒，冬日花园里的踱步，Spa 中的午后给人的感觉除了惬意，还有放松。

身边无处不是当地的瑰宝。登上“圣心”教堂（Sacre Coeur），便是巴黎的美景。萨尔瓦多·达利，就在不远处。乘上地铁，眼观巴黎的风景，聆听巴黎的声音。夜幕降临的时候，走出酒店，随性探访世界有名的美食酒店，或者准备在“红磨坊”度过一个令人兴奋的夜晚。

Travel back in time to the Belle Epoque at Hotel Maison Souquet – an exclusive Parisian townhouse in buzzing Montmartre. This new boutique hotel captures the decadent glamour of bohemian Paris – when Montmartre drew artists and aristocrats, socialites and dancers, who would meet and mingle in the districts theatres, wine bars and private salons. At the turn of the 20th century, Paris' pleasure houses played second home to the city's socialites and aesthetes. Hotel Maison Souquet has reproduced the sumptuous

decor of such a place, just a stone's throw from Montmartre and its famous Moulin Rouge.

Escape the hubbub of Paris for this atmospheric retreat, where each room exudes opulence. Take in the grand central staircase and study, decorated in Empire and nineteenth-century Orientalist style where intricate paneling and cordova leather furnishings complement eclectic ornaments and original artwork. Sip cocktails in the lounge, stroll around the winter garden or indulge in an afternoon of pampering at the Spa.

Hotel Maison Souquet puts many of Montmartre's treasures right at your feet. Climb up to the Sacre Coeur for exceptional views across Paris, or while away an afternoon at the nearby Salvador Dali gallery. Or you could hop on the Metro to explore the sights and sounds of central Paris. As the evening draws in, seek out world-famous fine dining restaurants, or enjoy an evening of cabaret and Can-can at the iconic Moulin Rouge.

传统新古典

TRADITIONAL
NEOCLASSICISM

T英伦骑士心
The Heart of British Knight

- 项目名称：洛阳英伦骑士心——紫悦府 B 户型别墅
- 设计公司：深圳市昊泽空间设计有限公司
- 设计师：韩松
- 用材：木饰面、大理石、石材马赛克
- 面积：600 m²

- Project Name: Unit B Villa of Pink-Pleasure Mansion
- Design Company: Shenzhen Horizon Space Design
- Designer: Han Song
- Materials: Veneer, Marble, Marble Mosaic
- Area: 600 m²

这个世界如果没有理想主义，人生还有什么意义，我们整天抱怨物欲横流，却也心安理得的沦陷其中。总是梦想着别人是否会蹦出来成为那个可以粉身碎骨的超级英雄，却从来没想过自己是不是可以成为任性一把的堂吉诃德。

我们心中持续向往的骑士，他优雅而粗犷，谦虚温和又孤傲勇敢；外表理性严谨，内心狂野不羁，感情用事，为了理想和原则却也可以放下执念与贪念……

我们今日缺失的，将来迟早要补上。

Without idealism, life would be meaningless and of no any significance. Even with all-day complaint on material desire, man feels nothing but easy and obsessed with possessions, always dreaming that some day someone can be a super hero and fighter devoted to an access to purity of land and never making any attempt to be a Don Quixote.

The knight we admire is a person who is gentle and rough, mild and brave, and reasonable in appearance and logic in mind, who is wild and unruly, but never acts impetuously, and who can give up greediness and obstinacy because of his ideals and principles.

What we are short of is bound to be compensated soon or later.

My box
My box

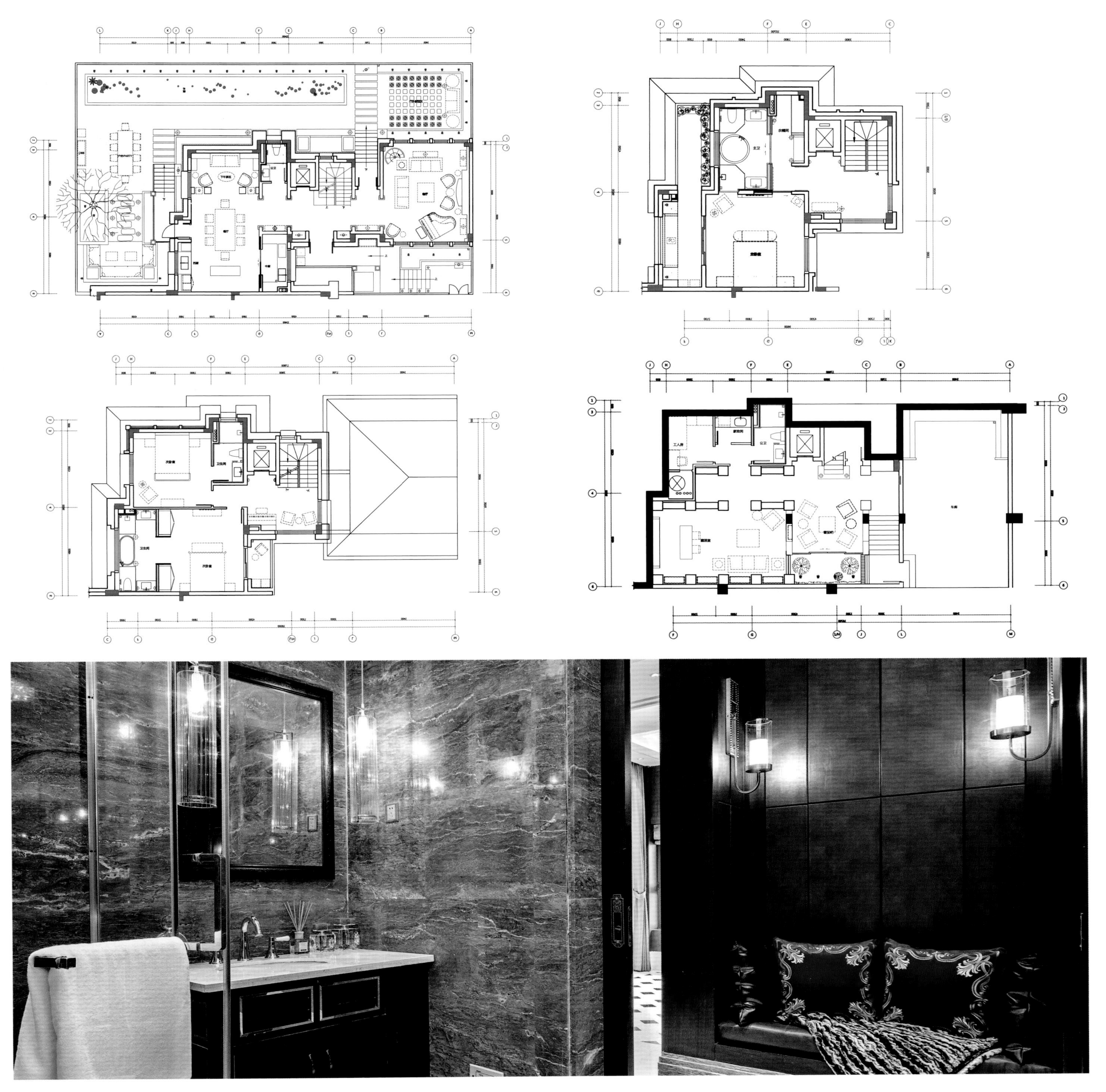

豪宅殿堂里的帝国梦想

An Imperial Dream in a Mansion

■ 项目名称：巴黎旺多姆威斯汀酒店

■ Project Name: The Westin Paris–Vendôme

巴黎旺多姆旁的威斯汀酒店，毗邻着协和广场，正对着巴黎旧王宫公园，好不让人羡慕的一个所在。世界闻名的卢浮宫不过咫尺之遥。周围无不是世界著名的地标性建筑。巴黎的购物天堂奥诺雷市郊路几分钟便可到达。繁华、时尚的街区，尽在这里。

酒店初建于 1878 年，经久不衰，堪为独家经典。法兰西第二帝国时期，其更贵为“国家宝藏”。酒店曾多次经历修缮，最近的一次整修可追溯至 2008 年。虽然历经雨打、风吹，但风采依旧。庭院中的喷泉水中女仙的大理石雕像也依旧挺立。1878 年的玻璃屋顶依然熠熠生辉。拿破仑三世 1898—1919 年流放期间，皇后尤金妮亚下榻的房间，就位于正门入口旁，并刊文予以记载。维克多·雨果 79 岁、81 岁、83 岁举办生辰庆典的皇家沙龙还在一楼。诸如圣洛郎、莲娜丽姿、杜兰朵等历史名人举办秀场的地方如今成了最为时尚的房间。

酒店共有 440 间客房。巴黎同类奢华酒店中，本酒店拥有套房量最多。包括“第二帝国”在内，所有套房均可观巴黎城市天际。著名设计师西贝尔（Sybille de Margerie）翻新设计的房间，可谓极为精致与现代。西贝尔以历史为灵感，着色淡雅，如淡紫、灰褐，间以绿色、枣红色。所有房间大床“威斯汀天梦”全为威斯汀创新之作。该卧床特别受客人的欢迎，如今已在威斯汀网店面向消费者发售。

The Westin Paris – Vendôme has an enviable location in the heart of Paris just off the Place Vendôme, next to the Place de la Concorde and immediately opposite the spectacular Tuileries Gardens. Just a stone's throw from the Louvre, the hotel is surrounded by famous landmarks and is minutes away from the city's best shopping on the elegant Rue du Faubourg St-Honoré. This is the capital's busiest and most fashionable neighborhood.

When the hotel was first opened in 1878 it was celebrated as the most exclusive property of its kind and was regarded as a national treasure of France's Second Empire. Over the years, it has been renovated and restored many times, most recently in 2008. But important historic details remain: a courtyard fountain with a white marble water nymph, and a spectacular glass roof from 1878. This is where Empress Eugenie, wife of Napoleon III, took up residence during her husband's exile from 1898 to 1919, as commemorated by an inscription near the front entrance. This is also where Victor Hugo held three large banquets celebrating his 79th, 81st and 83rd birthdays in the lavish Imperial Salon on the ground floor. The most celebrated fashion houses such as Yves Saint-Laurent, Hanae Mori, Nina Ricci, Torrente, and Ungaro regularly hold their shows in the hotel's historic salons.

With 440 rooms, The Westin Paris - Vendôme has the largest number of suites of all the traditional Parisian luxury hotels in its category. From their Second Empire classicism, the majority of the suites at the hotel offer exceptional views of the capital- truly one of the best panoramas of the Paris skyline. The famous designer Sybille de Margerie refurbished the rooms in a sophisticated contemporary style. Taking inspiration from the history of the hotel, Sybille de Margerie created an ambience of renewal and relaxation with muted color schemes of plum–mauve to taupe, with splashes of light green and Bordeaux reds. Of course all the rooms at The Westin Paris - Vendôme are equipped with the sleep-inducing Westin Heavenly Bed, one of the great innovations developed by Westin. The bed has proved so popular with guests that the Westin Heavenly Bed is now for sale at the Westin online store.

SIX SENSES
BALANCING SENSES

伦敦上流社会的瑰丽剪影
The Magnificence Sketch of the Upper-Class in London

■ 项目名称：多尔切斯特酒店

■ Project Name: The Dorchester Hotel

“多尔切斯特酒店”是世界上最具标志性酒店之一，位于伦敦一处安静之所。无论置身于精致的客房、套房，还是环绕式 Spa 内，客人都可以享受米其林美食，沉溺于一流的下午茶之中。永恒的魅力令人回味无穷。

俯瞰着海德公园，此处是世界名流、政要、皇室、上流社会不可多得的选择。开阔、奢华、精致的房间装饰着特别定制的织物、古董家具，卧室温馨舒适。洁白的大理石浴室内据说有伦敦最深的浴缸。

As one of the world's most iconic hotels, The Dorchester is quite simply the place to be in London. Whether you are staying in one of the exquisite rooms and suites, unwinding in the Spa, enjoying Michelin-starred cuisine or indulging in an award-winning afternoon tea, you will experience the epitome of timeless glamour.

The Dorchester, superbly located in the centre of London over looking Hyde Park, is a favorite choice of celebrities, world leaders, royalty and high society. Spacious, luxurious and elegant, each room features specially commissioned fabrics, antique furniture and exceedingly comfortable beds, while our white marble bathrooms are said to feature the deepest baths in London.

巴黎奢华与荣耀的缩影

The Epitome of Luxury and Glory in Paris

■ 项目名称：巴黎雅典娜广场酒店

■ Project Name: Hotel Plaza Athenee

新近修复的“巴黎雅典娜广场酒店”可观埃菲尔铁塔的壮丽景观，是巴黎奢华的缩影。

“雅典娜酒店”位于时尚的核心地带——蒙田大道。自 1913 年开业以来，便云集各路名流，达官显贵。今天的酒店更受时尚人士的青睐。

现代精致的巴黎内饰，美丽的客房灵感源于高档时装及舞场。身居于舒适、奢华的空间内，远处便是埃菲尔铁塔，近处则是庭院里飘扬的红色遮雨篷。夏日里，这里成就着户外美食的惬意。冬季里，酒店则拥有自己的滑冰场。

重新设计的酒吧由著名的调酒师坐镇。鸡尾酒，独一无二。餐厅则由世界级的米其林星级大厨主管。端坐于令人眩目的水晶天花下，纵情于新式概念的美食之中。啤酒店别致、流行，流行爵士乐在这里夜夜尽情欢演。

Spa 现代、潇洒，于祥和之中装备着与众不同的设施。除了美甲，这儿还可以修脚。

The newly refurbished Hotel Plaza Athenee is the epitome of Parisian luxury with spectacular views of the Eiffel Tower.

Since 1913, Hotel Plaza Athenee has been welcoming guests delighted to stay on avenue Montaigne, the heart of haute couture. Today the hotel is still perfectly positioned for those wanting to explore the city's designer fashion boutiques.

Discover the hotel's subtly updated Parisian interiors and beautiful new event rooms, inspired by haute couture, including the beautiful new ballroom. Stay in luxurious comfort overlooking either the Eiffel Tower or the charming courtyard with its signature red awnings, perfect for al fresco summer dining and in winter features its own ice rink.

Bar au Plaza Athenee has been lavishly redesigned and with mixologist Thierry Hernandez at the helm you can expect exceptional cocktails. Alain Ducasse, world-renowned multiple Michelin-starred chef, oversees all the hotel kitchens. In his Alain Ducasse au Plaza Athenee restaurant experience a new culinary concept amidst a stunning ceiling of crystal. In contrast, the listed Art Deco Le Relais Plaza is a chic and ever-popular Parisian brasserie, which hosts popular jazz nights.

To indulge in total serenity, Europe's only Dior Institute is another hotel highlight, offering exclusive Spa treatments in a stylish setting including a specialist manicure and pedicure area.

英国王室认证的尊贵与显耀

Dignity and Prestige Classified by British Royal

■ 项目名称：伦敦丽兹酒店　■ Project Name: The Ritz in London

创建于 1906 年的丽兹酒店位于伦敦最时尚的 St. James 区，在英国皇家花园格林公园（Green Park）和历史悠久的皮卡迪里环道（Piccadilly）中间。作为全球最有名望的酒店之一，丽兹酒店是第一个得到威尔士亲王钦定并获得英国王室供货许可证（Royal Warrant）的酒店，为英国王室提供酒店及宴饮服务，荣耀至极。酒店共有 136 间客房、1 家餐厅和 1 家酒吧，每周五、周六还有豪华的烛光晚餐舞会。特色鲜明的 Ritz Color，也就是蓝色、桃红色、粉红色和黄色的色彩组合充满酒店每一个房间，24 克拉重的金叶和英国古典家具，路易十六时代的皇家气派一览无遗。从 1906 年开始，这间著名的酒店接待了无数的名人学者、皇亲贵族和明星名流。贵宾客人包括国王爱德华、女王伊丽莎白和她的母亲、丘吉尔等。

房内以暖色调酒店设计装饰，设有壁炉、高天花板和华丽吊灯。客房建筑特色令人惊叹，完美结合现代化设施与古董家具。

新古典主义建筑特色体现豪华水准，令人惊叹，内设豪华卧室，并为客人提供经典英式下午茶和精致美食。

Between the imperial park, Green Park and Piccadilly Circuit, The Ritz in London was established in 1906 in St. James, the most fashionable district. As one of the most eminent hotels, The Ritz was the first to be granted Royal Warrant by Prince Wales. The hotel has 136 guest rooms, one restaurant, one bar, and space for candlelight dinner on Fridays and Saturdays. Ritz color is the signature hue of the brand, consists of blue, peach-red, pink and yellow that are fused in each room. Out of the 24-karat gold leaf and classical British furnishings, flows the imperial magnificence that's exclusive to Louis XVI. Since its openness, numerous celebrities, families from English Family, stars and scholars have been accommodated, like King Award, Queen Elizabeth, Elizabeth's mother and Churchill.

Coated in warm hue, rooms feature in fresco, towering ceiling and chandelier, which with the combination of modern facilities and antique furniture raises eyebrows.

It's neo-classical but it's also of luxurious standard. When overlooking Green Park, here provides classical English afternoon tea and delicious foods.

RITZ

曾经显贵的殿堂 今日城市中心的绿洲

Yesterday's Palace of Dignitary, Today's Oasis in City Center

■ 项目名称：巴黎拉斐尔酒店

■ Project Name: Hotel Raphael, Pairs

巴黎就是这样一个地方，有着很多隐匿于都市里的角落。角落里另有绿洲，安静得让人有一种与世无争的感觉。巴黎拉斐尔酒店就位于这样一个角落。顶层餐厅周围绿树红花，远望是凯旋门、埃菲尔铁塔。古旧的内饰，彰显着富丽的丝绸，富有品位的古董，彰显着古巴黎风的锦缎墙纸。7层的建筑建于1925年，原是达官显贵的宫殿。1.8米高的窗户，宽大的大型衣橱还是曾经的模样。不远处，香榭丽舍大街步行便可以到达。祥和的家居氛围，给人一种大隐于市的感觉。亲密的空间里，当然不乏精致的服务。

Paris is full of hidden corners that offer calm oases, and one of the prettiest is found at Hotel Raphael. Raphael La Terrasse, the hotel's rooftop restaurant, is beautifully planted and overlooks the Arc de Triomphe and Eiffel Tower. Classic Paris is also present in the old-world interiors, which feature rich silks, tasteful antiques and damask wallpaper. Housed in a seven-storey palace that was built in 1925, the hotel preserved such original details as 1.8-meter-tall windows and large armoires (even if guests are no longer bringing along ball gowns). In walking distance to the Champs Elysées but feeling worlds away, the hotel fosters a serene, home-like vibe, thanks to its intimate size and discreet service.

让 230 年的荣光耀目重现

The Reemerged Halo of 230 Years Ago

■ 项目名称：波尔多大酒店 Spa

■ Project Name: Grand Hotel de Bordeaux& Spa

波尔多金三角的核心地带，大剧院的对面，便是“波尔多大酒店 Spa”。位于酒店中心的后部地带，Spa 美学修复之后，给人一种与众不同的享受。翻阅历史，便不难发现“大酒店”对当地的重要意义。自 18 世纪末期酒店开业，至 230 年后的今天，空间虽经兴衰，但对于城市的生活、当地的历史遗产、当地的葡萄园，“大酒店”都极其重要。不同的人生相逢与感悟在此都可以上演。

1999 年，当地商人米歇尔买下这片房产及其相邻建筑，并决意使“波尔多大酒店”荣光重现。8 年后，商人的目标得以实现。曾经的入口地带如今华丽转身，成了 Spa 所在。更加宏伟的空间，雅克·加西亚的大师之作，给人极其深刻的印象。大理石，当地石头，紫红、淡紫、深红等暖色调的织物，18 世纪风格的家具，原创花艺给人一种宁静、安详的感觉。客房、套房同样追求雅致，而精于细节。和谐与纯粹尽在空间之中。

In the heart of the golden triangle, opposite the Grand Théâtre, the Grand Hôtel de Bordeaux & Spa is back on centre stage as an exceptional product after an authentic restoration. When we look at this building's history we understand why the Grand Hotel is so important to the people of Bordeaux. The Grand Hôtel de Bordeaux & Spa has become part of the life of its city, its heritage, its region and its vineyards. What was true from the end of the 18th century, at the very beginning of its history, is still true 230 years later. The Grand Hotel de Bordeaux & Spa is the place for all types of meetings and for all emotions.

In 1999, Bordeaux businessman Michel Ohayon decided to buy this facade and the adjacent buildings with the firm intention of reviving the myth of the Grand Hotel. Eight years later, he had achieved his goal, and at the end of 2007 the doors of what is now called the Grand Hotel de Bordeaux & Spa opened revealing a world more majestic than ever, brilliantly orchestrated by designer Jacques Garcia. The majestic hall of the Grand Hôtel de Bordeaux & Spa designed by Jacques Garcia is impressive. Marble, Bordeaux stone, warm-tone fabrics (plum, mauve, deep red), 18th century-inspired furnishings, original floral creations: an invitation to peace and tranquility. The suites and rooms are in the same spirit, they are places of calm and luxury. Each detail has been given great attention in the purest refinement. You will get a sensation of serenity and comfort thanks to this hushed wellbeing.

Bordeaux

穿越时空，体验文艺复兴欧洲古堡宫殿

Go Through Time Tunnel to Experience European Castle in Renaissance

- 项目名称：大连城堡豪华精选酒店
- 设计公司：HBA、姜峰设计
- 面积：80 000 m^2

- Project Name: The Castle Hotel, a Luxury Collection Hotel Dalian
- Design Company: HBA, Jiang & Associates
- Area: 80,000 m^2

城堡，一个充满欧洲浪漫气息的名词，在大连童话般地呈现。

大连城堡豪华精选酒店坐落于大连市莲花山山脉，俯瞰著名的星海湾和黄海，是大连市的地标性建筑。酒店位于亚洲最大的城市广场星海广场，紧靠国际会展中心、贝壳博物馆以及风景优美的海岸线，是游客最喜爱的滨海路的起点。该项目由一方集团开发，喜达屋酒店运营，众多国际顶尖团队协作，加上 J&A 的参与，早已决定了大连城堡豪华精选酒店的高端定位，设计上的精雕细琢，与大连城堡豪华精选酒店在星海湾畔迸发出经典的共鸣。

酒店的外观保留了原有欧式风格，其典雅外墙由手工精心挑选的石块筑就。这座富丽堂皇的巴伐利亚城堡式的特别建筑已被 CNN 推荐为“全球十大邦德主题酒店”。走入大连城堡豪华精选酒店，我们仿佛穿越时空来到了追求极致城堡艺术的古代欧洲，从巨大的油画、精美的铁艺扶手到立体的大理石装饰，宛如文艺复兴鼎盛时期的古堡宫殿。

大堂的穹顶是由半透明的玻璃建成，这样就可以让自然光柔和地洒向大堂。玻璃上雕刻着精美的图案，形成了三棱镜的反射效果。阳光通过三棱镜，在大堂的地面上留下了一道道小小的彩虹。大堂中的楼梯扶手，为了打造出暗黑色的视觉效果，设计师们在扶手的下面镀上一层金，再在金上镀铜，经过几番调整才达到最佳效果。

酒店拥有 292 间客房与套房（包括一间超过 600 平方米开放式布局的总统套房）及 67 间公寓，或临海或傍山。内部装潢精美，海洋景观迷人。顶级团队设计改造后的城堡，呈现豪华而现代的风格，从墙面装饰、家具选择到艺术收藏，每一个设计细节背后都藏着一个故事。站在海景房的窗边，看着夕阳的余晖映照在星海湾上，正应了那句诗：你在桥上看风景，看风景的人在楼上看你。

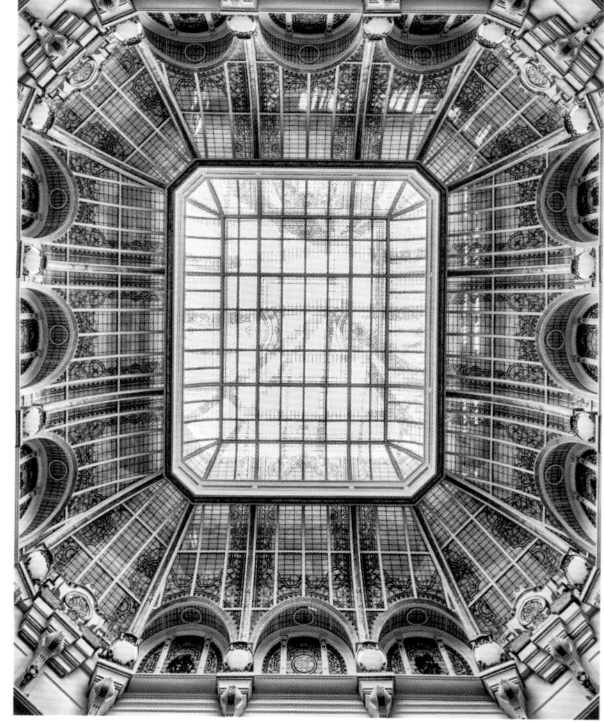

大连城堡豪华精选酒店拥有三家风格各异的餐厅，提供创意美食和地道佳肴。臻宝餐厅秉承“海洋到餐桌”和“农场到餐桌”的赏味理念，提供最原生态的海鲜和菜品；全日餐厅集锦餐厅的菜单则呈现多样化环球美馔；皇室啤酒坊拥有自酿德国鲜啤和北欧甄选美食。此外，宾客还可在行政酒廊度过闲适时光：在典雅的大堂吧悠然享用传统英式下午茶，抑或在华丽的贵裔廊细细品味干邑白兰地和红酒。

酒店的地理位置优越，是举办会议和活动的理想之地。酒店活动场地包括 1 间大宴会厅、7 个多功能厅以及面积达 4 000 平方米的露天屋顶花园，花园风格大气典雅，适合举行盛大婚礼、私人鸡尾酒会和重要商务会议等活动，见证各种重要时刻。

酒店宽敞的室内泳池、设施一流的健身中心和奢华的凯撒水疗中心，让客人的入住体验更为丰富。健身中心设有一间多功能瑜伽室，室内温水泳池则拥有一窗壮观的海洋景色。凯撒水疗中心为渴望放松的客人提供了一方休憩的绿洲，其中的桑拿浴室和全面身体护理项目可以帮助宾客驱散身心疲惫，还有 3 间可用于聚会社交的娱乐室。

大连城堡豪华精选酒店，一个让童话再现，美梦成真的地方。

Castle, a word rich in European romance, now appears in Dalian in a fairyland manner.

The Castle Hotel is seated at the Lianhua Mountain in Dalian, overlooking the Xinghai Bay and the Yellow Sea, another landmark of the city. Located at the largest city plaza in Asia-Xinghai Plaza and adjacent to the International Convention Center, the Shell Museum and the coastline with gorgeous scenery, it is also the starting point of the most beloved Binhai Road by tourists. Developed by Yifang Group, operated by Starwood Hotel, and supported by a variety of world-class teams, including J&A, it's certainly high-end positioned, while its exquisite design makes resonance with its physical construction in the bay.

Its facade of carefully-selected stone is kept its European style. The gorgeous castle has now listed as Top Ten Bond Theme Hotel in the World by CNN, where you feel as if you were going through time tunnel into European in ancient times when to seek an utmost castle art, and whose huge oil paintings, beautiful iron handrail, and solid marble decoration make nothing but an old castle in the prosperous time of Renaissance.

The lobby dome is of transparent glass for sunlight to come. On the glass is exquisite pattern to have prim's reflection effect, so that sunlight makes rainbow one after another. The stairs' railing is glided below and then coated with copper to allow for dark visual effect.

Including one presidential suite with an area over 600sq.m, all 292 guest rooms and suits and 67 apartments are either by sea or hill, each decorated finely and beautifully. The refurbished space into castle looks luxurious and modern, where each detail from wall decoration, furnishing and art collection has its own story. Views by window reaches into the bay, where the sun is setting, just like the description in one poem, that you enjoy landscape on bridge, while watched over by people who seek scenery upstairs.

Of all three dining restaurants, each is differentiated, where to house cuisine and authentic dishes. The Utmost Treasure offers the most ecological seafood and dished when sticking to an idea from the sea and the farmer onto the table. The all-day one has a global menu. The Imperial Beer provides self-brewed fresh German beer and catering from North Europe. The Excusive lounge is right here for you to idle your time away. The lobby features traditional British afternoon tea. And the Noble Lounger is for

brandy and red wine.
Here in all makes a good place for conference and activity with its location. Besides the large banquet hall, 7 multi-functional halls and the roof garden measuring as large as 4,000 square meters, its garden of grandeur and generosity is good for large wedding party, private cocktail and important business activities.
With the interior swimming pool, the state-of-art fitting center and the Spa, the experience is thus enriched. In the fitting center, there is a yoga room. The pool enjoys a broad view that comes from the sea. The Spa has a relaxing oasis and three recreation rooms besides sauna and all-body care service.
Here it is, the Castle Hotel in Dalian, a place where to reappear the fairy tale and to realize your good dream.

昔日英国国王的专属酒店再现动人华彩

Hotel Ever Exclusive to British King, the Resplendence at Current

■ 项目名称：皇家依云度假村

■ Project Name: Hotel Royal - Evian Resort

“皇家依云度假村”开业于 1909 年 6 月 6 日，获世界最美酒店之美誉。酒店名称归功于英国国王爱德华七世。当时有一套房专为其预留。室内设计由建筑师艾伯特（Jean-Albert Hébrard）完成。经典的造型如白色的巨轮。空间白色的线条融“新艺术派”、“装饰艺术”风格为一体。公共空间的拱顶、圆形大厅以淳朴风格的壁画作为装饰，呈现出一派水粉画的世界。百年魅力，见证无数国际大事件的发生。世界各地的名人政要曾云集于此。2003 年是法国总统选举年。同年 6 月， G8 峰会于此召开。时任法国总统希拉克在此会见了下榻于此的各国首席代表。

2014 年 7 月，修缮一新的酒店再次开业。整修后的设计为了呈现昔日经典的荣光，完全尊重了初建时的设计。同时，最新款式的现代设计糅合于其中。二楼的客房，虽然改变了风格，却很优雅、和谐。从酒吧至大厅，从酒窖至图书室，新巴洛克式的壁画已经恢复了往昔的光彩。穹顶、拱顶变得更为优雅美丽。150 个客房、套房中，一半以上在得到恢复的同时，更为现代。红木、紫檀、锡兰柠檬木材质的家具添了铜绿，专为曾经的那一份沧桑。六楼至今还在使用的会务空间、奢华私人套房得到了更新升级。波光粼粼的水景伴着阿尔卑斯山的巍峨达到了新的境地与高度。底楼新开张的会务中心，俯瞰着花园。Les Fresques 餐厅，皮质、木质家具的直线条以及铜质吊灯升华着壁画曲面的笔触。旧时空间的魔力谱写出新的诗篇。

The Hotel Royal was inaugurated on 6 June 1909. Dubbed the "most beautiful hotel in the world" at the time, it owes its name to the King of the United Kingdom, Edward VII, for whom a suite was immediately reserved. Designed by the architect Jean-Albert Hébrard, its classical form recalls a vast white liner. Its clean lines blend Art Nouveau and Art Deco styles. The vaults and rotunda of the shared spaces are brightened with rustic frescos in watercolours.

In this environment favorable to agreements, for around a century, numerous international encounters have taken place, and the greatest heads of state from all over the world have stayed. In June 2003, year of the French presidency, Evian hosted the G8. Chief delegates were welcomed by Jacques Chirac on the terrace of the Hôtel Royal where they were staying.

The renovation of the Hôtel Royal, was reopened in July 2014, is a large-scale project. Respecting the original design, the aim is to restore this centennial hotel to its former luminous glory. Blending historical heritage with the latest trends in contemporary design, the first- floor rooms have been restyled in elegant harmony. From the bar to the great hall, the wine cellar to the library, Gustave Louis Jaulmes's neo-Baroque frescos that grace the building's vaults and domes have regained their former luster. Half of the hotel's 150 bedrooms and suites have been restored and modernized. The delicate patina on the period furniture in precious woods – mahogany, rosewood, Ceylon lemonwood – specially designed for their original setting, has been restored. On the sixth floor, hitherto intended for meetings and conferences, luxurious private suites have been converted, whose view of the lake and the Alps reaches new heights. On the ground floor overlooking the garden, a new conference centre has been opened. In the Les Fresques restaurant, the straight lines of the leather and blondwood furniture and the lighting from the brass chandeliers compliment the curved strokes of the frescos and reinvent the place's original magic.

殿堂级的尊贵与精致

The Palace-Level Honor and Delicacy

■ 项目名称：佩斯塔纳宫殿酒店暨国家纪念馆

■ Project Name: Pestana Palace Hotel & National Monument

经过精心的修复，“佩斯塔纳宫殿酒店”于美丽的环境中款款地向大众走来。奢华、精致的酒店有美丽的“塔霍河”为伴。如今的酒店可是“世界潮流酒店”的特权成员，同时被列举为葡萄牙国家里程碑。丰富的私家园林簇拥左右。视线穿过亚热带的绿植与树木，是里斯本全景式的景观。

优越的位置、完善的设施、极好的品质，高标准的服务为其赢得了不同团体颁发的奖项。难以超越的精致美食又为其在业内赢得了几项国际大奖。

Set in the truly beautiful surroundings of the carefully restored 19th century palace Valle-Flôr, the Pestana Palace Hotel welcomes visitors. This luxurious and elegant hotel has wonderful views to the Tagus River, is a privileged member of the 'Leading Hotels of the World' group and is listed as a Portuguese National Landmark.

An exuberant private park surrounds the grounds with an abundance of subtropical plants and trees presenting a magnificent panoramic vista over the city of Lisbon.

Its wonderful location and facilities, excellent quality, standards and service are only some of the reasons why the Pestana Palace Hotel has received so many prestigious awards from different independent associations. The Hotel has also won several international competitions for their unsurpassed, fine dining.

T 意大利文艺复兴艺术的辉煌呈现

The Glorious Presence of the Renascence

■ 项目名称：意大利依葵莎洲际大酒店

■ Project Name: Niquesa's Grand Hotel Continental

“依葵莎洲际大酒店”所在古城锡耶纳，是中世纪小城，是联合国教科文组织文化遗产。从酒店出发，几步之外便是铺满鹅卵石的小道，那可是中世纪的遗物。不远处，便是历史文化古建筑，如著名的“坎波广”、“哥特式罗马天主教堂”。可以说，酒店的建筑是意大利文艺复兴时期建筑的典范。宽大的房间、旧时的壁炉、价值连城的艺品、古董……17 世纪初建时，是某贵族的府邸。

包括客房、套房在内，共有 51 个房间。古董家具、名贵织物、旧时的布局在空间翻新时得到极为精心的呵护，一如初见。穿过古典式玻璃门，便进入玄关大厅。高高的壁画天顶，18 世纪风格的“灯笼”式照明两相辉映。前院的右边是现代的 SaporDivino 酒吧餐厅。新加的透明天顶使其如同一个冬日的花园一般。两个现代的玻璃钢构电梯直通三楼，升华着经典建筑、石质艺品带给人的体验。

美丽的古典式房间给人一种丰富、奢华的感觉。大量的罕见古董、绘画、丝绸、瓷器、大理石、洞石等丰富着人的视觉体验。

套房位居两层塔楼的顶端，可谓是名副其实的奢华天地，远看城市天际及周边的群山、峡谷，隐私感依然独好。无论是有木横梁的客厅，还是宁静的书房，都拥有全景式的俯瞰景观。卧房位于柱承式阳台之中。整个套房除了琳琅满目的古董，便是优雅、和美的气氛与环境。墙面覆盖着珍贵的丝绸。浴室里，柔性的大理石，温和舒适。沐浴于美丽的环境中，不经意间，便是历史的回响。雅致的卧室，已静静地闪耀了几百年。拥有精湛手艺的大师们以其艺术灵感在此诠释着空间的定义。

Siena's exquisite Centro Storico is the medieval heart of this ancient city and a UNESCO World Heritage site. The Grand Hotel Continental is steps away from Siena's cobbled medieval lanes and the city's stunning architectural features, including the famous Piazza del Campo and the Romanesque - Gothic cathedral. The boutique hotel is a paragon of fine Renaissance architecture, with opulent rooms boasting original frescoes, priceless artworks and precious antiquities. It was built in the 17th century as an aristocratic residence.

The original interior layout was retained in creating 51 rooms and suites furnished with antique furniture and costly fabrics, where painstaking care is lavished on every detail. A classic glass revolving door provides access to the magnificent entrance hall, where the high frescoed ceilings of Palazzo Gori Pannilini are lit by lantern-shaped lamps in the eighteenth - century style. A pleasant surprise is provided on the right by the contemporary SaporDivino wine bar & restaurant located in the former courtyard, now covered by a transparent dome and converted into a winter garden. The two modern elevators of glass

and steel serving the three floors complement the classic architecture and stone work.

The beautiful classic rooms are unashamedly opulent. Your stay will be enhanced by a luxurious environment and enriched by a curated collection of precious antiques and paintings; and finishes of silk, porcelain, marble and travertine.

Perched at the very top of the palazzo in the two-storey tower, the Royal Suite is a veritable kingdom of luxury and refined privacy with incomparable views over the city and its surrounding hills and valleys. From the parlor with its wood-beamed ceiling to the quiet study with its panoramic views of the world below and the bedroom on an altana, all of the rooms of this magnificent suite create refined and harmonic ambiences graced with antiques. The walls are covered in precious silk and bathrooms are made with soft cool marble to invoke historical inspiration in this beautiful setting. The hotel's elegant hotel bedrooms have been honed over centuries; defined by master craftsmen who have embellished them with their art.

T 阿姆斯特丹的不朽传奇

The Legend of Amsterdam

■ 项目名称：阿姆斯特丹欧洲旅馆　　■ Project Name: De L'Europe Amsterdam

“欧洲旅馆”被人们尊为阿姆斯特丹传奇旅馆。25 年来，该酒店一直是“世界潮流”的成员。酒店位于阿姆斯特河岸，历史街区核心地带。19 世纪时期的瑰宝空间，底蕴丰富。永恒的设计，亲密的气氛，慷慨的服务尽在本案。

整个空间的装饰与众不同。每个客房与套房都以荷兰大师的画作为主打，这些画作原本被阿姆斯特丹国家博物馆所收藏。111 个美丽房间，23 个奢华套房位于侧翼“荷兰大师楼”，另外 88 个客房、套房位于 Rondeel 大楼。该酒店拥有阿姆斯特丹唯一一个配备有 6 个卧室的套房。Rondeel 大楼是酒店最初的建筑，位于城市绿洲，如今成了优雅的缩影。此处的每一间客房、套房无不是经典的设计。iPad、地暖、移动性地灯更给人一种家居般的享受。大胆的用色，丝绒的地毯，硬木的地板，定制的织物，精致的麻线，墙面上荷兰大师灵感的画作统一于和谐的空间中。

现代的“荷兰大师楼”和谐的不仅有艺术，还有设计。中庭的天际线，令人眩目的艺品升华着精品酒店的通透感。“荷兰大师楼”里的 23 个套房，大胆、开阔。里面的画作，皆为大师手笔。原作则陈列在附近的国家博物馆。套房或气氛亲密，或一居室设计，但无不配备着令人舒适的设施，无不运用最新科技。

De L'Europe is acknowledged as Amsterdam's legendary hotel and is a member of The Leading Hotels of the World for over 25 years.

With its location on the banks of the Amstel River in the heart of historic, it's impossible to escape the city's rich history while staying at this iconic 19th century treasure. Expect timeless design, intimate ambience and genuine service.

De L'Europe boasts a distinctive decor, enhancing each room and suite with a replicated Dutch Master painting, handpicked in exclusive partnership with the nearby Rijksmuseum.

De L'Europe showcases 111 beautifully appointed guestrooms, including 23 luxurious suites in the Dutch Masters Wing and 88 rooms and suites in the Rondeel Building. De L'Europe is the only venue in Amsterdam boasting a six bedroom Signature Suite.

An oasis of calm in Amsterdam's city centre, De L'Europe's original Rondeel Building epitomises timeless refinement. With each guestroom and suite incorporating classic design elements and cutting-edge comforts such as iPads, heated flooring and motion-detecting floor lighting, the Rondeel presents a homely, welcoming environment. Bold colours combine with plush carpeting, hardwood floors, custom fabrics and fine linens, while

Dutch Masters-inspired paintings adorn the walls to create a sense of restful refinement.

The contemporary Dutch Masters Wing harmonises art and design, with its atrium skyline and dazzling works of art creating the effect of an airy boutique hotel. Each of the 23 boldly appointed, spacious suites displays a different Dutch Masters painting recreated from the original in the nearby Rijksmuseum. Guests may select an intimate suite or a spacious one-bedroom suite, each featuring indulgent amenities and the latest technology.

图书在版编目（CIP）数据

浪漫新古典Ⅵ / 黄滢，马勇 主编．– 武汉：华中科技大学出版社，2015.9

ISBN 978-7-5680-1274-4

Ⅰ．①浪… Ⅱ．①黄… ②马… Ⅲ．①住宅 – 室内装饰设计 – 图集 Ⅳ．① TU241-64

中国版本图书馆 CIP 数据核字（2015）第 242445 号

浪漫新古典Ⅵ　　黄滢 马勇 主编

出版发行：华中科技大学出版社（中国·武汉）
地　　址：武汉市武昌珞喻路 1037 号（邮编：430074）
出 版 人：阮海洪

责任编辑：熊纯　　责任监印：张贵君
责任校对：岑千秀　　装帧设计：筑美文化

印　　刷：利丰雅高印刷（深圳）有限公司
开　　本：889 mm × 1194 mm　1/12
印　　张：23
字　　数：138 千字
版　　次：2016 年 1 月第 1 版 第 1 次印刷
定　　价：358.00 元（USD 71.99）

投稿热线：（020）36218949　　duanyy@hustp.com
本书若有印装质量问题，请向出版社营销中心调换
全国免费服务热线：400-6679-118 竭诚为您服务